D0247889

# SIGMA 2

## INVESTIGATING MATHEMATICS

Keith Hedger and David Kent

with David Burghes and Don Steward

HODDER AND STOUGHTON

LONDON   SYDNEY   AUCKLAND   TORONTO

**Other books in the SIGMA series:**
SIGMA 1 Investigating Mathematics
SIGMA 1 Teacher's Book
SIGMA 2 Teacher's Book

Kent, David
  SIGMA: investigating mathematics.
  2
  1. Mathematics – For schools
  I. Title  II. Hedger, Keith
  510

ISBN 0 340 39759 4
First published 1988

Copyright © 1988 D. Burghes, K. Hedger, D. Kent and
D. Steward

No part of this publication may be reproduced or
transmitted in any form or by any means, electronically or
mechanically, including photocopy, recording, or any
information storage or retrieval system, without either the
prior permission in writing from the publisher or a licence,
permitting restricted copying, issued by the Copyright
Licensing Agency, 33–34 Alfred Place, London WC1E 7DP.

Phototypeset by Gecko Limited, Bicester, Oxon
Printed in Great Britain
for Hodder and Stoughton Educational
a division of Hodder and Stoughton Ltd, Mill Road
Dunton Green, Sevenoaks, Kent by
Butler & Tanner Ltd, Frome and London

# Contents

# Preface

This book is part of the SIGMA Project, **Studies and Investigations into Generating Mathematical Activities**, which aims to enhance mathematical teaching and learning in school.

The project involves a range of activities, the aim of which is to provide material and suggest a method of approach which enriches the mathematics curriculum, provokes discussion and the use of a wide range of equipment and resources.

SIGMA 1 is primarily intended for the 11 to 14 year old age group but some of the material may be appropriate for older and younger pupils. Similarly SIGMA 2 is primarily intended for the 14 to 16 year old age group but its use need not be restricted to this age range. The Teacher's Books which cover SIGMA 1 and SIGMA 2 indicates clearly our suggested approach for using the material, and it is expected that staff and pupils will wish to extend topics by adding their own material and ideas.

The common core activities take a particular mathematical theme and give ideas for motivating pupils and developing the topic. The investigations introduce, develop and give possible extensions for particular mathematical problems, whilst the applications develop the use of mathematical analysis applied to practically based problems.

In all of the activities we encourage the teaching of mathematics through discussion, investigative techniques, applications and problem solving — alongside the traditional exposition and practice of routines. The activities have been written so that they can be readily incorporated into an existing mathematics curriculum, and SIGMA 2 has topics which can be incorporated into a GCSE coursework assessment scheme for any examining board.

The project was originally financed by the Southern Regional Examination Board.

Project Directors:

David Burghes    Professor of Education, University of Exeter

Keith Hedger    Mathematics Adviser, Shropshire LEA

David Kent    Mathematics Adviser, Tameside LEA
(formerly Mathematics Adviser, Suffolk LEA)

Don Steward    Head of Mathematics, Oldbury Wells School, Bridgnorth, Shropshire

## Note for teachers

Many of the exercises contained in this book can be assessed as school-based GCSE Mathematics Courseworks. The Teacher's Book gives details of a recommended methodological approach together with a suggested mark-scheme where appropriate.

They are intended to act as occasional components of a departmental 4th and 5th year curriculum and as such are not arranged in any particular order in the book.

## Note for students

### Coursework

If you have to present any of the chapters in this book as part of your coursework, remember the following points which will help you to write a proper report.

Include in the work:
● any methods or strategies you used for the work
● any observations you made either during the work or from the results
● a record of all the results from your investigation
● an analysis of your results
● any generalisations you can make about your results
● if possible, proof of the generalisation
● any extensions to the work that you carried out.

# 1 Flip Over 1

## Four cards

A playing card can lie either face up  or face down.

Lay out 4 cards in a square, all face down:

You can **flip over** either a whole row, or a whole column. So if you flip over the first column you get the following finishing pattern:

## Question 1

What finishing patterns do you get if you:

 (i) flip over the second column then flip over the first row

(ii) flip over the first column then flip over the first row?

Illustrate your answers with diagrams.

6

## Question 2

What combination of flips give the finishing patterns:

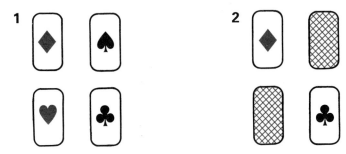

You might think that a certain finishing pattern can be obtained in more than one way. If so, give as many ways as you can.

## Question 3

How many different finishing patterns can you make? Illustrate your answer.

## Nine cards

Now take 9 cards, all laid face down and in a square. We will have the same rule that either a whole row or a whole column is flipped over at any one time. So if we flip over the second row we get:

## Question 4

What finishing patterns do you get if you:

 (i) flip over the third column, then flip over the second row

(ii) flip over the third row, then flip over the first column, then flip over the second row?

Record your results by drawing the patterns.

## Question 5

Which flip or combination of flips give:

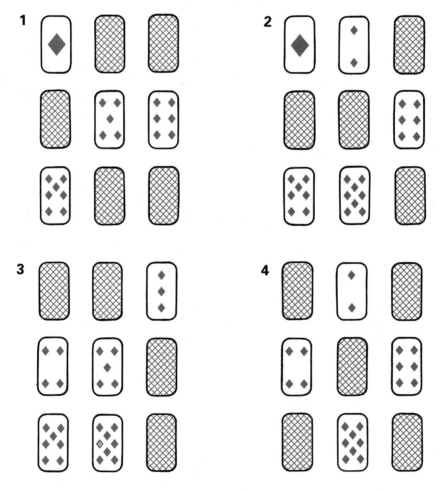

Give as many combinations as you can for each pattern.

# Question 6

How many different finishing patterns can you make with the 9 cards in a square?

# Question 7

You have looked at the number of finishing patterns for a 2 × 2 and 3 × 3 square. Can you find a formula for the total number of finishing patterns for an $n \times n$ square?
If you can find such a **generalisation**, prove it.

# Question 8

Is this finishing pattern one of those that you have made?
If it is, say what combination of flips you used to make it. If it is not, **prove** that it cannot be made.

# 2 Flip Over 2

We shall introduce some symbols to represent row and column flips.

$C_1$ means flip over the 1st column. $R_2$ means flip over the 2nd row.
$C_1 \bullet R_2$ means flip over the 1st column then flip over the 2nd row.

We can now write down, in symbols, an **algebra** of the flips.

## Question 1

Investigate this algebra using the following points as a guide to your work.

(i) Is the operation $\bullet$ **commutative** (does $a \bullet b = b \bullet a$ — always?)

(ii) Is the operation $\bullet$ **associative** does $a \bullet (b \bullet c) = (a \bullet b) \bullet c$?

(iii) Is there an **identity element, I** (such that $a \bullet I = a$)?

(iv) Solve equations such as $C_2 \bullet R_1 \bullet y = R_2 \bullet R_3 \bullet C_1$ (using 9 cards).

(v) Make some of your own equations and swap them with your neighbours.

(vi) Simplify expressions such as $C_1^3 \bullet R_2^2$.

(vii) Create some expressions to simplify. Exchange these with your neighbours.

## Question 2

What happens to the results of the Flip Over 1 investigation if you allow **diagonal** flips?

(i) How does it alter the algebraic results if we have **D₁** and **D₂** flips?

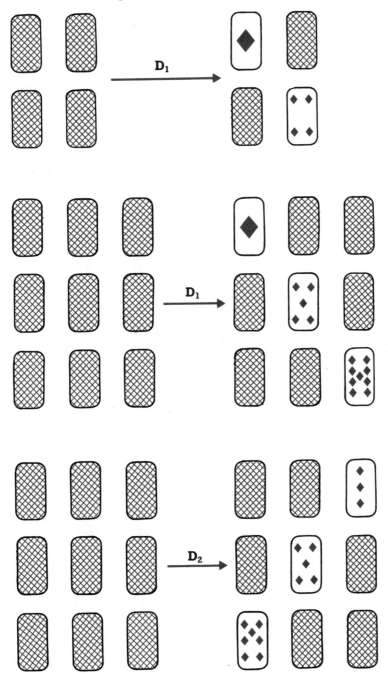

(ii) Can you express $D_1$ in terms of **C** and **R** flips?

(iii) How many finishing patterns can you obtain using **D**, **C** and **R** flips?

(iv) Is this pattern one of them?

## Question 3

Suppose we lay out 9 dice in a square grid like this:
again we only have row and column flips — that is we must do the same flip to all the dice in a particular row or column.

Your task is to investigate this situation. In particular look at:

  (i) the various finishing patterns

 (ii) the algebra of the flips

(iii) whether or not it is possible to obtain the finishing pattern shown on the right and if it is possible, prove it can be done

(iv) what happens when we include diagonal flips.

12

# 3 Pythagoras 1

## Question 1

Work out the areas of squares *A, B* and *C* in each case and compare (area *A* + area *B*) with (area *C*). As you record your results note whether the triangle is (a) an acute-angled triangle (b) a right-angled triangle (c) an obtuse-angled triangle.

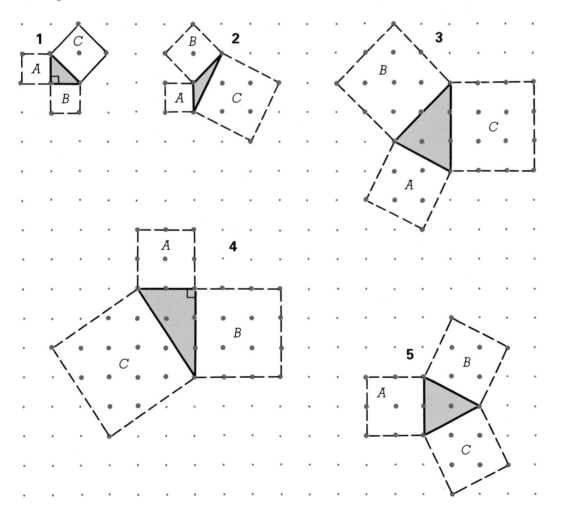

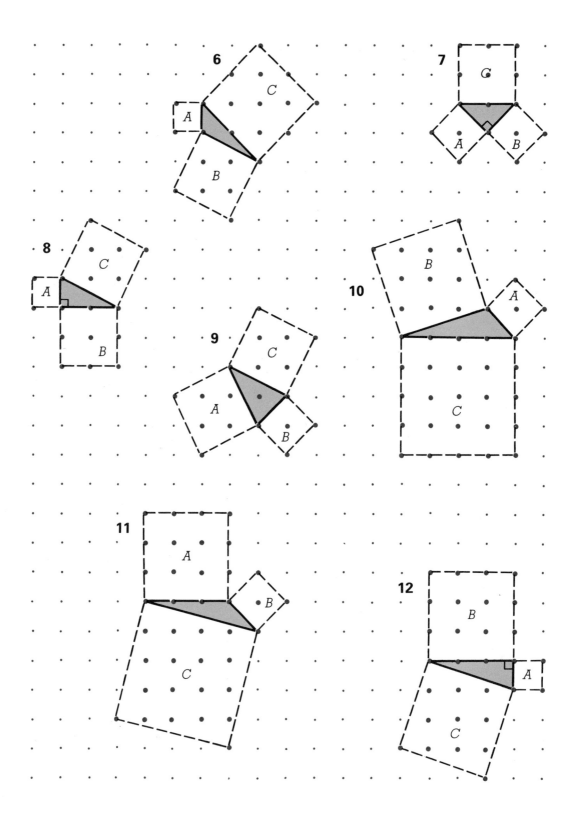

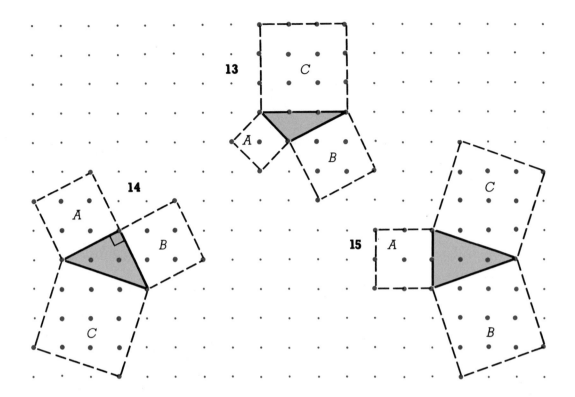

Write down what you have noticed about your results

# Question 2

From the special result for a right-angled triangle, you should be able to work out a rule that allows you to find the areas marked $x$ in the next question. All the lengths shown are in centimetres.

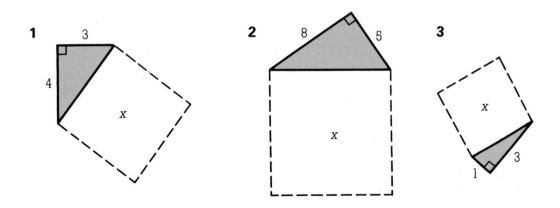

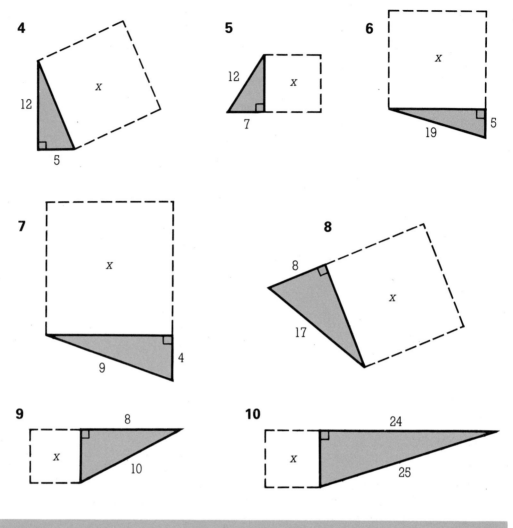

## Pythagoras

The rule you have been using is known as Pythagoras' Theorem. It is usually written

$$a^2 + b^2 = c^2$$

where $a$, $b$ and $c$ are the side lengths of a right-angled triangle. $c$ is always the longest side. The theorem is named after a greek mathematician called Pythagoras who lived from 580 BC to about 500 BC.

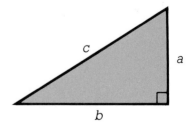

# Question 3

Use the theorem to find the missing side length in the following diagrams.

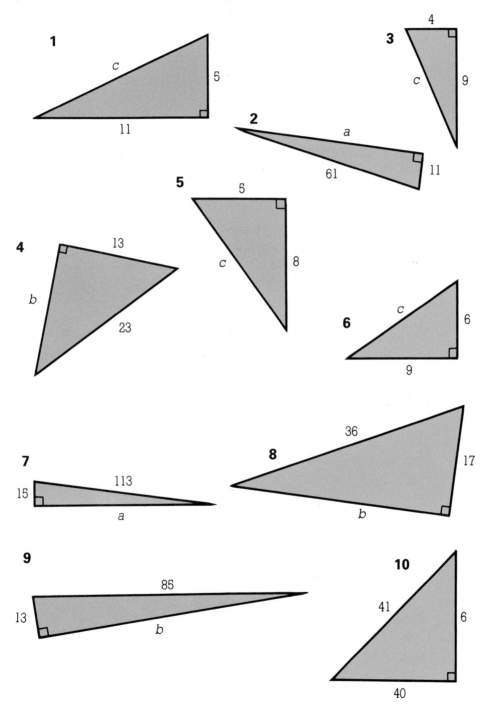

**1**

c

5

11

**2**

a

61

11

**3**

4

c

9

**5**

5

c

8

**4**

13

b

23

**6**

c

6

9

**7**

113

15

a

**8**

36

17

b

**9**

85

13

b

**10**

41

6

40

## Question 4

Find out the lengths of these lines using Pythagoras' Theorem. You will need a calculator. Arrange them in order of size, smallest to largest.

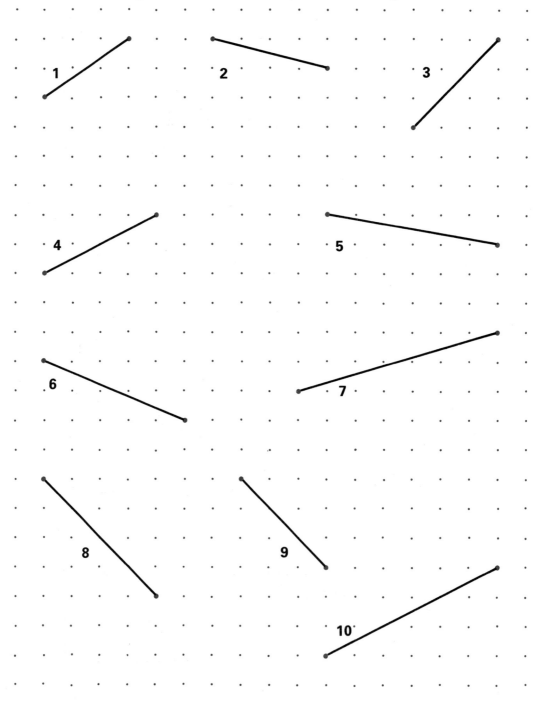

## Question 5

How many different lengths of line
can you find on this 3 × 3 lattice?

## Question 6

Draw several different triangles on a
3 × 3 lattice. Use a calculator to work
out the perimeter of the triangles.

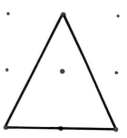

# 4 Routes on Polyhedra

The rules are:

(i) always start at a top vertex (•) and finish at the bottom vertex (X) which is diagonally opposite the start

(ii) move only along the edges

(iii) you cannot move upwards on the vertical edges

(iv) you cannot move along an edge more than once on any particular route

(v) you cannot pass through the finish point.

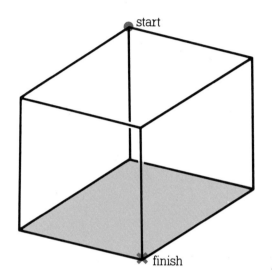

## Question 1

What is the shortest route from start to finish? What is the longest route?
How many of each are there?
How many different routes are there from the start to the finish?
Show all your working in detail, including diagrams of the routes you took.

## Question 2

Do the same for each of the following prisms:

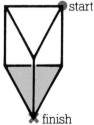  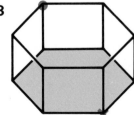

# Question 3

Construct a table of values for the prisms like this:

| number of edges on base | total number of routes |
|:---:|:---:|
| 3<br>4<br>5 | |

# Question 4

Can you see from your work and from the table, what the **general relationship** connecting the number of routes with the number of edges on the prism base, might be?

# Question 5

What happens to the number of routes in the case where the finishing point is *directly below* the start?

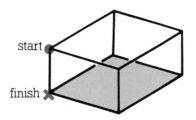

# Question 6

What are the results for these pyramids?

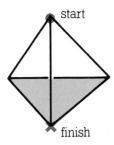
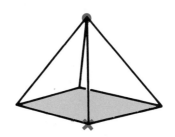
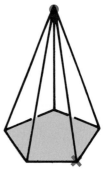

Record your results as for question 2.

What is the **general relationship** for pyramids?

## Question 7

Can you *prove* or give some sort of detailed *explanation* as to *why* the results for the prisms and the pyramids are as they are?

## Question 8

What is the result for the cube if we keep all the rules but relax the one about not moving upwards? What happens now for the other prisms? Can you offer a **general relationship** in this case?

## Question 9

What happens when a pyramid is placed on top of an appropriate prism.

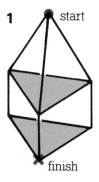

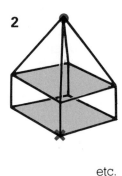

Can you offer:

(i) a demonstration of the results in the cases above

(ii) a **general relationship**

(iii) a proof or some sort of explanation of the general result (in words and with diagrams perhaps)?

## Question 10

What happens when 2 pyramids are placed base to base:

Can you offer:

(i) a **general relationship**

(ii) a proof or explanation of the generalised result?

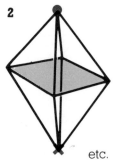

## Question 11

Can you see any way in which this problem can be extended? Could you write a computer program which:

 (i)  traces the routes on a cube

 (ii) gives you the required number of routes no matter how many sides there are on the base of the pyramid or prism

(iii) allows you to examine the routes when frameworks are combined?

# 5 Cuisenaire Activity 1

Here is a reminder of the different Cuisenaire rods available. *White* is 1 unit length and each following rod increases in length by 1 further unit.

w : white
r : red
g : light green
p : pink
y : yellow
d : dark green
b : black
t : tan (brown)
B : blue
O : orange

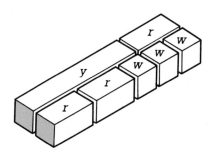

## Simple equations

## Question 1

Answer the following in terms of a single rod:

1. $g + r + y = \square$
2. $y + g - d = \square$
3. $t - \square + w = b$
4. $(y + r) - g = \square$
5. $y - (r + \square) = r$
6. $\square - (b + r) = w$
7. $w + g + g + \square = B$
8. $y + r = 2r + w + \square$
9. $B - (2r + p) = \square$
10. $w + \square + y = b$
11. $d + (b - 2g) = \square$
12. $2w + r + \square = y$
13. $\square = O - (2r + g)$
14. $b - (w + \square + g) = r$
15. $3y - 2p = \square$
16. $y - 4\square = w$

**17** $3y - 2(r + w) = \square$

**18** $3y - 2r = \square + 2w$

**19** $3y - \square = 2w + 2r$

**20** $r - \frac{1}{2}\square + p = g$

## Cuisenaire fractions

## Question 2

Find a single rod as an answer to these:

**1** $\frac{1}{4} \times t =$

**2** $\frac{1}{5} \times O =$

**3** $\frac{1}{5}(O + y) =$

**4** $\frac{1}{4}(2t) =$

**5** $\frac{1}{3}(2d) =$

**6** $\frac{1}{4}(t + p) =$

**7** $\frac{1}{5}(3y + O) =$

**8** $\frac{1}{7}(5t + r) =$

**9** $\frac{2}{3} \times B =$

**10** $\frac{3}{5}(O + y) =$

**11** $\frac{5}{6}(O + r) =$

**12** $\frac{3}{8}(O + d) =$

**13** $\frac{3}{4} \times (\frac{2}{5} \times O) =$

**14** $\frac{5}{6} \times (\frac{6}{7} \times b) =$

**15** $\frac{3}{4} \times (\frac{8}{9} \times B) =$

**16** $\frac{2}{3} \times (\frac{2}{3} \times B) =$

**17** $\frac{3}{5} \times (\frac{1}{2} \times \square) = g$

**18** $\frac{4}{5} \times (\frac{3}{4} \times \square) = d$

**19** $\frac{1}{4} \times (\frac{2}{3} \times (\frac{6}{7} \times (\frac{7}{9} \times \square))) =$

**20** $\frac{3}{4} \times (\frac{2}{3} \times (\frac{6}{7} \times (\frac{7}{9} \times (\frac{3}{5} \times 3\,\square)))) = g$

## Further Equations

## Question 3

Solve these equations using a single rod for each missing element. If there are two missing elements, still only one rod is required.

**1** $2\square = d$

**2** $2\square + g = b$

**3** $2\square - p = d$

**4** $3\square - t = O$

**5** $5\square + p = B$

**6** $4\square - B = b$

**7** $3\square + y = O + b$

**8** $4\square + p = O + d$

**9** $\square + r = 2\square - r$

**10** $\square + g = 2\square - r$

**11** $3\square - r = \square + p$

**12** $3\square = \square + t$

## True or false?

## Question 4

Use the rods to see if the following statements are true or false. Copy each one down and write true or false beside it. For each one that is false, try to write a correct version changing as little as possible of the statement.

1  $r + g = g + r$

2  $(w + p) + g = w + (p + g)$

3  $2(g + w) = 2g + w$

4  $y - r = r - y$

5  $r + (y - p) = (r + y) - p$

6  $O - (y + p) = O - y + p$

7  $B - (r + w) = B - r - w$

8  $(w + O) - y = w + (O - y)$

9  $B - 2r = B - r + r$

10  $(b + y) - p = b + (y - p)$

11  $O - 2g = (O - g) - g$

12  $(t - p) - w = t - (p - w)$

13  $B - (g + y) = B - g + y$

14  $3y - 2p = (2y - p) + (y - p)$

15  $O - 2(r + w) = O - 2r - w$

16  $5g - 3p = (3g - 2p) + (2g - p)$

17  $5g - 3p = (4g - 3p) + g$

18  $5g - 3p = (5g - p) - 2p$

19  $5g - 3p = (6g - 3p) - g$

20  $5g - 3p = (6g - 3p) - g$

Now try to write $6p - 4y$ in at least 5 different ways.

## Fraction equations

## Question 5

Find the rod that replaces □.

1  $\left(\dfrac{2\square + p}{5}\right) + g = y$

2  $\left(\dfrac{3\square - g}{3}\right) - r = w$

3  $\dfrac{\left(\dfrac{2\square - t}{2} - r\right)}{5} + w = r$

4  $\dfrac{\left(\dfrac{7\square - 2p}{2}\right) - r}{4} = r$

5  $\dfrac{\frac{1}{2}(3\square + 2p)}{4} - r = r$

6  $\dfrac{\frac{5}{6}\square + b}{3} - w = g$

7  $\dfrac{2p - \frac{1}{3}\square + r}{3} + w = p$

# 6 Cable Television

Cable TV provides a local TV service with many channels. One of the problems that a new company has to solve is that of wiring up the connections between local towns in the most economical way.

For example, suppose we want to connect up all the towns around Sheffield using the routes shown in the diagram. All the distances in this diagram and others in the chapter are in kilometres. We could use the connection shown in black in the second diagram:

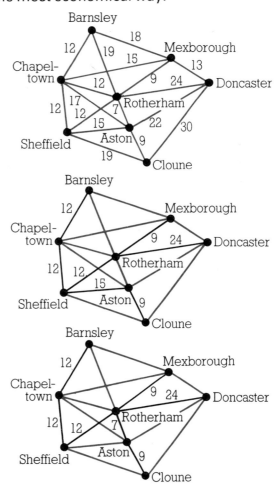

The total number of kilometres connecting cable is

$$12 + 12 + 12 + 9 + 24 + 15 + 9 = 93 \text{ kilometres}$$

But can you do better? Can you find another connection between all the towns which is shorter? It's easy to see that joining Aston to Rotherham rather than Sheffield will reduce the total by 8 kilometres. So we now have the connection shown in the third diagram. This gives a total of 85 kilometres.

## Question 1

Find a connection between all the towns which uses only 74 kilometres of cable.

With simple networks like the one above it is usually easy to find the shortest connection by trial and error. But when the network gets more complex, it's not quite so easy. Try to find the minimum connection in the next three questions.

## Question 2

Find the shortest connection between all the towns shown in the diagram.

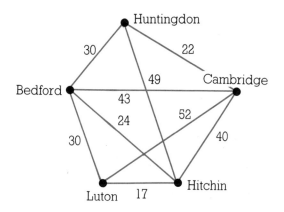

## Question 3

Can you use this table to draw a sketch map showing how each place can be connected to a system of roads so that the total road length is as short as possible?

Derby

| 132 | Hull | | | | |
|-----|------|------|-----------|----------|------|
| 105 | 82 | Leeds | | | |
| 88 | 142 | 60 | Manchester | | |
| 55 | 97 | 49 | 57 | Sheffield | |
| 57 | 55 | 36 | 90 | 78 | York |

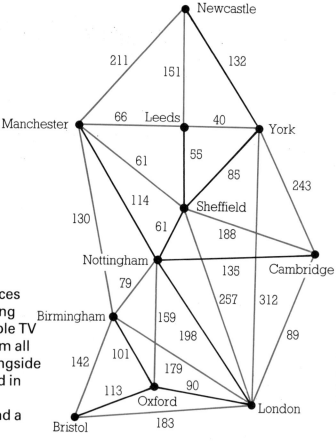

## Question 4

This diagram shows the distances between a number of cities along main roads. A new national cable TV company wants to connect them all to London by laying cables alongside main roads. The route indicated in black on the map uses 1065 kilometres of cable. Can you find a layout which uses less?

Having tried these examples, you have probably found a way of finding the shortest connecting path. The best way is to start at any town (or city) and then follow the rule:

Find the next nearest town to those already connected.

Keep on repeating this rule until all the towns have been connected. How this system works for question 4 is shown below.

| | | Total length |
|---|---|---|
| *Step 1*<br>Start with any city,<br>e.g. London | ● Lo | 0 |
| *Step 2*<br>Join London to the nearest<br>city, Cambridge | Lo● 89 ● Ca | 89 |
| *Step 3*<br>Join the next nearest city to<br>either London or Cambridge<br>which is Oxford | Ox● 90 Lo 89 ● Ca | 179 |
| *Step 4*<br>Join the next nearest city to<br>London, Oxford or<br>Cambridge, which is<br>Birmingham | Bi●<br>101<br>Ox● 90 Lo 89 ● Ca | 179 |

Continue this process for yourself. It should take 11 steps and you should finish at Newcastle.

## Question 5

Using this method, find the minimum connection for the network shown in this diagram:

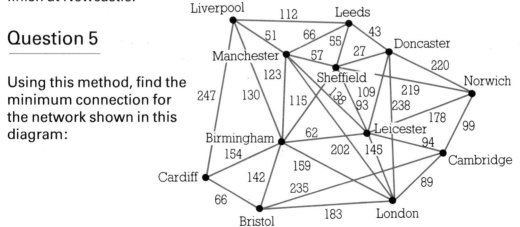

# 7 Rearrange the Formula

## Example

This is an example of a loop diagram. Whatever happens to the number or letter as it moves from the left-hand box to the right-hand box, the inverse must happen to convert back to the original as it moves from right to left. Therefore we complete the box like this:

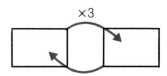

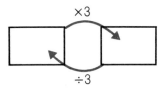

Put some numbers in the left-hand box and follow them through the loop.

## Question 1

Copy and complete the following loop diagrams:

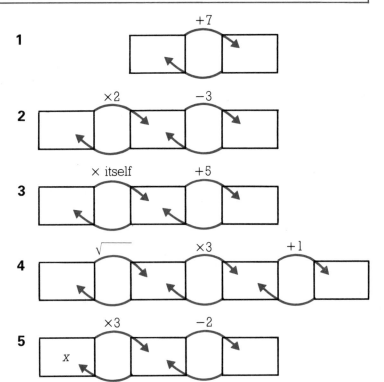

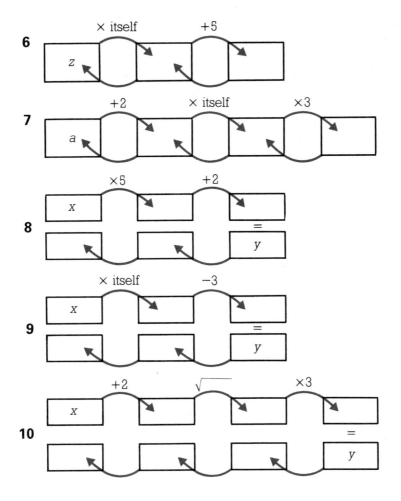

**6** $z$    × itself    +5

**7** $a$    +2    × itself    ×3

**8** $x$    ×5    +2    = $y$

**9** $x$    × itself    −3    = $y$

**10** $x$    +2    $\sqrt{\phantom{x}}$    ×3    = $y$

# Question 2

In 1–6 on the next page make the letter in brackets on the right the subject of the formula on the left. You may want to do them using the method shown in the example or you could use a method of your own.

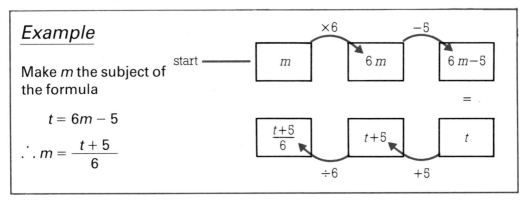

*Example*

Make $m$ the subject of the formula

$$t = 6m - 5$$

$$\therefore m = \frac{t + 5}{6}$$

start  —   $m$    ×6    $6m$    −5    $6m-5$

$\frac{t+5}{6}$    ÷6    $t+5$    +5    $t$    =

**1** $y = 5x - 7$     $(x)$        **4** $y = \sqrt{x}$     $(x)$

**2** $p = 3r + 1$     $(r)$        **5** $v^2 = 2fs$     $(s)$

**3** $s = at + b$     $(t)$        **6** $m = r^2 + 2$     $(r)$

## Question 3

Make $x$ the subject of each of the following equations:

**1** $y = ax^2 - b$            **4** $y - x = b$

**2** $y = mx$            **5** $y = a(x - b)^2$

**3** $y = \dfrac{1}{x}$            **6** $y = a - bx$

## Question 4

Rearrange the formula to make its subject the letter shown in brackets.

**1** $A = LB$     $(B)$        **6** $\dfrac{1}{v} + \dfrac{1}{u} = \dfrac{1}{f}$     $(f)$

**2** $s = at^2 + b$     $(t)$        **7** $s = ut + \frac{1}{2}gt^2$     $(g)$

**3** $m = a\sqrt{n} - b$     $(n)$        **8** $c = \frac{5}{9}(f - 32)$     $(f)$

**4** $s = \dfrac{a + b}{2}$     $(a)$        **9** $A = k(R^2 - r^2)$     $(R)$

**5** $s = \dfrac{(u - v)t}{2}$     $(v)$        **10** $T = 2k\left(\sqrt{\dfrac{n}{g}}\right)$     $(g)$

## Question 5

Given that     $T = k\left(\sqrt{\dfrac{1 - x}{1 + x}}\right)$

(i) calculate $T$ if $x = 0.1$ and $k = 5$

(ii) calculate $x$ if $T = \dfrac{k}{10}$

## Question 6

(i) If $V = \pi r^2 h$ and $A = 2\pi rh$, find $V$ in terms of $A$ and $r$.

(ii) If $P = \dfrac{mv - mu}{t}$ and $P = mf$, find $v$ in terms of $u$, $f$ and $t$ only.

(iii) Given that $a = \dfrac{3k + 5m}{k + 3}$, express $k$ in terms of $m$ and $a$.

## Question 7

Given that $z^2 = x^2 + 3xy$, calculate

(i) $z$ when $x = 24$ and $y = 42$

(ii) $y$ when $x = 25$ and $z = 55$.

# 8 Squares Investigation

A geoboard may be useful for this investigation.

## Question 1

Using only the lattice points on your paper, or geoboard, copy the diagram below and complete the square in each case. Note that two *adjacent corners* are shown. Then make up some of your own.

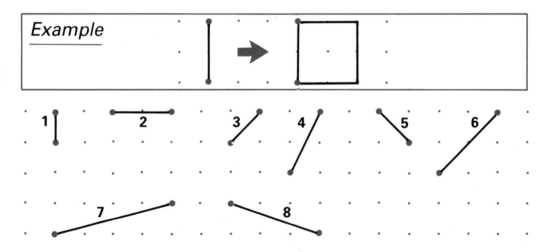

## Question 2

This time a corner is marked with • and the centre of the square is marked with **x**. Complete the square.

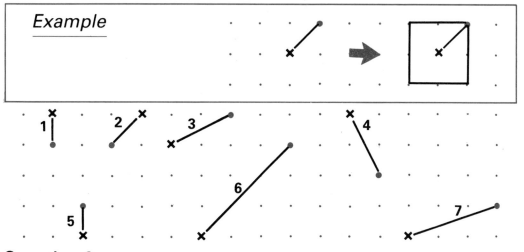

*Example*

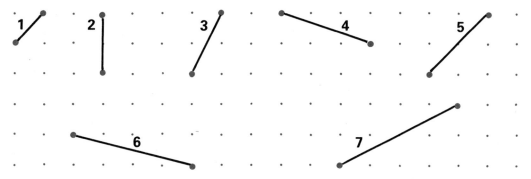

## Question 3

In the following cases the opposite corners are shown. However, in some you will find it impossible to make the square.

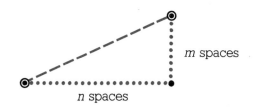

## Question 4

By looking at further cases, can you tell when the other corners of the square will be *on* lattice points and when *not*?

In general, when the opposite corners are shown, what rule connecting $n$ and $m$ determines when the other corner points of the square are on lattice points?

*m* spaces

*n* spaces

## Question 5

There are 6 different squares on the (3 × 3) or 9-pin board. How many
squares are on there on the (4 × 4), (5 × 5) . . . (n × n) board?

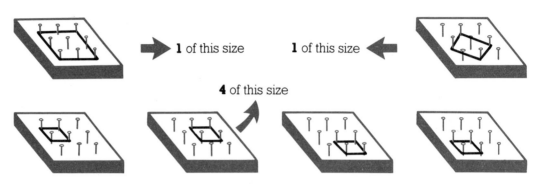

## Question 6

There are 9 rectangles in total for a (2 × 2) square. Investigate a **general rule**
for any (n × n) square.

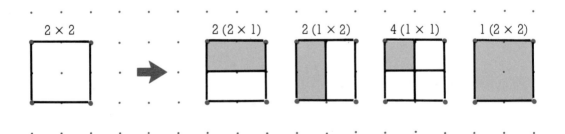

## Question 7

There are 11 squares in a (2 × 4) rectangle. These are shown in the diagrams
at the top of the next page. Investigate a **general rule** for any (n × m)
rectangle.

1(2 × 4)
8 (1 × 1)

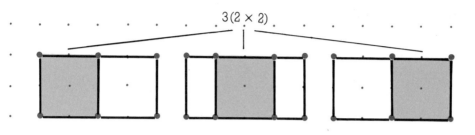

3(2 × 2)

## Question 8

There are 8 (1 × 1 × 1) cubes and 1 (2 × 2 × 2) cube, giving a total of 9 cubes in a (2 × 2 × 2) cube.

Use isometric dot paper or multi-link cubes or your imagination to investigate a general rule for the number of cubes in any ($n \times n \times n$) cube.

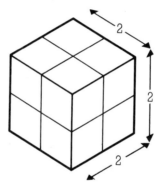

# 9 Graphical Transform 1

## Straight lines graphs

All these graphs derive from the equation $y = x$.

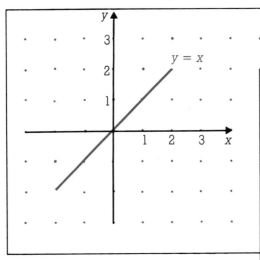

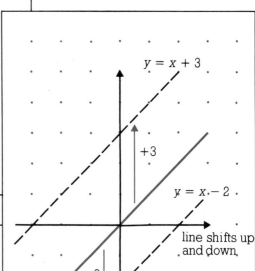

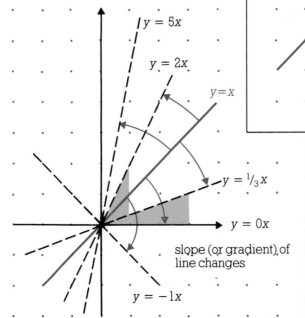

Adding and subtracting terms shifts the line up or down. Multiplying $x$ by a factor changes the slope or **gradient** of the line.

## Example

Sketch $y = 4x + 3$ showing each stage.

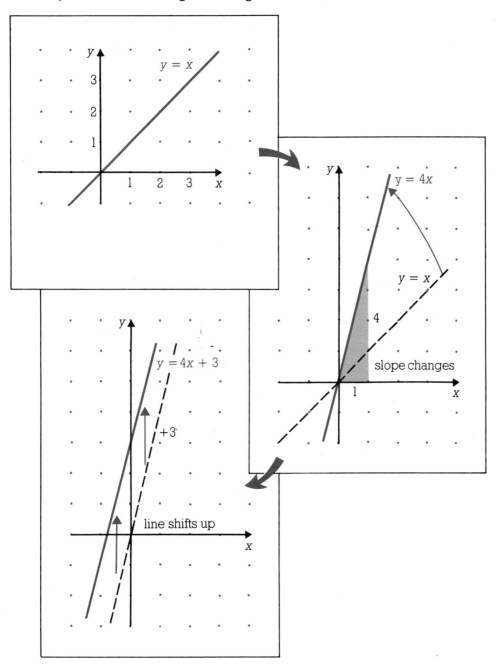

# Question 1

Sketch the following graphs. Draw separate quick sketches for the different stages like those shown on the last page.

**1** $y = 2x$          **6** $y = -25x$

**2** $y = 5x$          **7** $y = \frac{3}{4}x + 1$

**3** $y = 25x$          **8** $y = 2x + 1$

**4** $y = -x$          **9** $y = 17x + 3$

**5** $y = -2x$          **10** $y = 2x + 9$

## *Example*

Sketch

$$2y + 3x = 5$$

First rearrange the equation (see chapter 6 in this book).

$$2y = -3x + 5$$
$$y = -\tfrac{3}{2}x + \tfrac{5}{2}$$

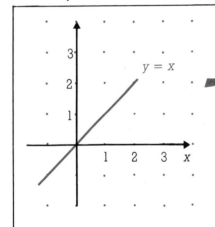

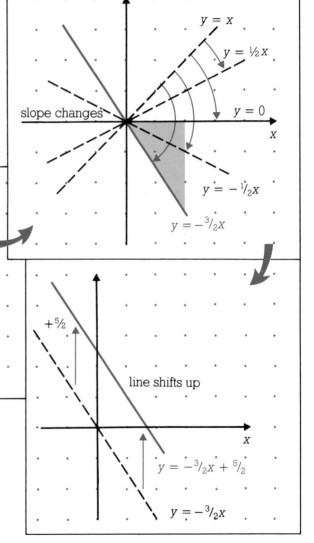

## Question 2

Sketch the following, showing the different stages which begin at $y = x$ in each case:

1  $y = 2x + 1$

2  $y = 5x + 3$

3  $y = x + \frac{3}{4}$

4  $y = 17x + 3$

5  $y = 2x + 9$

6  $y = 8 - \frac{3}{4}x$

7  $y - x = 2$

8  $2y = 4x + 6$

9  $4y - 12x = 20$

10  $3y + 4x = 12$

## Question 3

Copy and write down the equations of the following lines (show all your working) using the principles of transformation if possible.

1
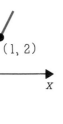

2
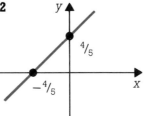

3
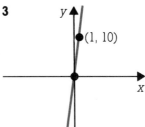

4
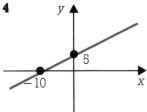

5
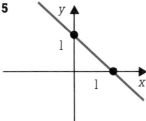

6
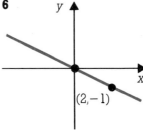

7
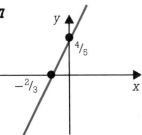

8
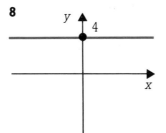

9
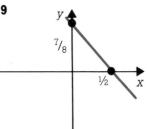

41

# 10 Cuisenaire Activity 2

w : white
r : red
g : light green
p : pink
y : yellow
d : dark green
b : black
t : tan (brown)
B : blue
O : orange

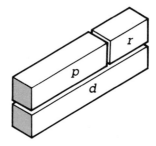 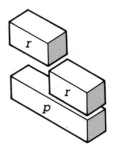

## Simultaneous equations

*Example*  What 2 rods add up to the O and subtract to make the p?

$$L + S = O$$

| O | |
|---|---|
| L | S |

$$L - S = p$$

L : longer rod
S : shorter rod

## Question 1

Now try these questions. Draw diagrams if it helps you. Try to find a systematic method for solving this type of question.

**1** L + S = B⎫
     L − S = g⎭

**2** L + S = t ⎫
     L − S = r ⎭

**3** L + S = d⎫
     L − S = p⎭

**4** L + S = b⎫
     L − S = g⎭

**5** $L + S = O$
  $L - S = d$

**6** $L + S = B$
  $L - S = w$

**7** $L + S = d$
  $L - S = r$

**8** $L + S = B$
  $L - S = y$

**9** $L + S = O$
  $L - S = p$

**10** $L + 2(S) = t$
   $L - \quad S = r$

## Question 2

In the following simultaneous equations, L = long rod, M = medium rod and S = short rod.

**1** $L + 2S = O + g$
  $L - \quad S = b$

**2** $2L + S = O + p$
  $L - S = w$

**3** $2L + \quad S = O + g$
  $3L + \quad 2S = O + t$

**4** $2L + 3S = O + b$
  $L + 2S = O$

**5** $3L + 3S = O + t$
  $L + 4S = O + r$

**6** $2L + 5S = 2O + b$
  $4L - 2S = 2O + r$

**7** $3L + 4S = 3O + d$
  $L - 2S = r$

**8** $2L + 3S = 2O + w$
  $3L + 2S = 2O + p$

**9** $L + 4S = O + d$
  $4L + \quad S = 3O + p$

**10** $3S - \quad L = b$
   $2L + \quad S = O + p$

**11** $3S - 2L = g$
   $L + \quad S = O + g$

**12** $L - 2S = O + b$
   $3S - 2L = w$

**13** $S + M + L = B$
   $3S - M + L = y$
   $3S + M - L = w$

**14** $S + \quad M + \quad L = O + b$
   $3S + 2M + \quad L = 2O + t$
   $2S + \quad M + 2L = 2O + b$

**15** $L + \quad M + 2S = O + p$
   $M + 2L - \quad S = O + t$
   $S + 2M + 3L = 3O + p$

**16** $2L - 3M + \quad S = b$
   $3L - 2M - 2S = O$
   $M + 3S - \quad L = y$

**17** $L + 2M + 3S = O + b$
   $3M - \quad L + 5S = O + g$
   $2L - \quad M + 4S = O + t$

**18** $2L + 3M - 6S = B$
   $L - 2M + 2S = r$
   $L + \quad M - 3S = r$

## Rod statements

*Example*

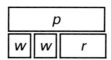

Here are some statements that can be made about these rods:

$$2w + r = p$$
$$p - w = r + w \qquad \text{Can you write at least 5 more?}$$

## Question 3

Write as many statements as possible for each of the following arrangements.

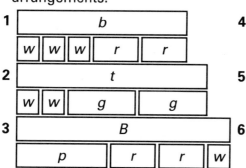

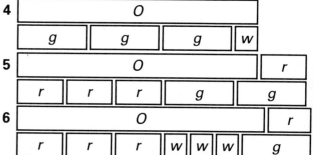

## Rearrangements

*Example:* Look at the diagram at the top of the page.

$$p - w = r + w$$
$$r = ?$$

The answer is:

$$r = p - w - w$$
$$\text{or} \quad r = p - 2w$$

## Question 4

Set up rods representing the rod equations shown and then write each equation in terms of the rods indicated.

**1** $w + t = y + p$

$w = ? \; t = ? \; y = ? \; p = ?$

**2** $O - g = y + r$

$O = ? \; y = ? \; r = ? \; g = ?$

**3** $O - (g + r) = p + w$

$O = ? \, p = ? \, w = ? \, p = ? \, (g + r) = ?$
$r = ?$

**4** $2r + b = g + t$

$b = ? \, g = ? \, t = ? \, 2r = ? \, r = ?$

**5** $O - (g + w) = y + w$

$O = ? \, y = ? \, g = ? \, w = ?$

**6** $3r + 2g + p = O + d$

$p = ? \, O = ? \, d = ? \, g = ? \, r = ?$

## Removing brackets

You should be able to see that the following are true:

(i)  $2(g + w)$  =  $2g$  $+ 2w$

So $2(g + w) = 2g + 2w$ and the brackets have been removed.

(ii) $3(p - r) = 3p - 3r$

(iii) $p - (r + w) = p - r - w$

(iv) $g - (r - w) = g - r + w$

## Question 5

Remove the brackets from these statements. Use the rods to check your answers.

**1** $3(2w + r) =$

**2** $5(p - w) =$

**3** $3(2g - p) =$

**4** $2(3g + p) =$

**5** $3(O - 2g) =$

**6** $t - (g + r) =$

**7** $B - (y - r) =$

**8** $b - (2r + w) =$

**9** $y - (p - g) =$

**10** $B - (2r + g - y) =$

# 11 Passola

This piece of work involves a ball game called **Passola**. It would be best to form yourself into groups and play the game.

## 5 player Passola

The 5 players stand in a ring.
A passes the ball to B, then B to C, then C to D, etc.

This version of Passola is called 'step 1' because you pass the ball to the person right next to you. So our diagram on the right shows 5 player Passola step 1.

The lower diagram on the right shows 5 player Passola step 2 because you pass the ball not the next person, as in step 1, but to the second person.

The diagram below shows 8 player Passola step 2. Some players are missed out altogether!

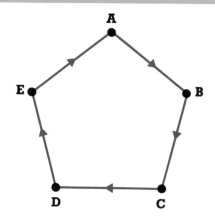

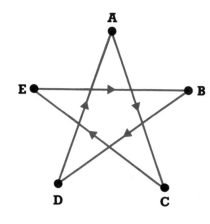

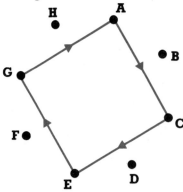

46

This diagram shows 8 player Passola step 3. Everyone receives the ball.

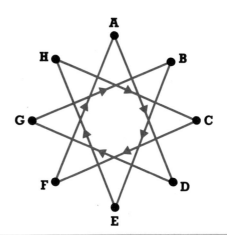

Your task is two-fold. Sometimes all players receive the ball and sometimes players are missed out. For part 1, find out when and why this is the case.

For part 2, find a rule to determine whether or not a particular player will receive the ball.

## Question 1

Try the Passola game with a number of players varying from 4 to 12 and at steps varying from 1 to 11. Record your results to show whether all people receive the ball or not.

## Question 2

State whether or not all people receive the ball for

**1**  6 player Passola, step 4

**2** 15 player Passola, step 6

**3** 15 player Passola, step 7

**4** 25 player Passola, step 10

**5** 25 player Passola, step 6

**6** 37 player Passola, step 6

## Question 3

Try to offer a generalised result which will predict whether or not all players receive the ball in the game of Passola.

## Part 2

Look at the diagram for 8 player Passola step 3. Starting at **A**, the path of the ball can be described as:

$$A \rightarrow D \rightarrow G \rightarrow B \rightarrow E \rightarrow H \rightarrow C \rightarrow F \rightarrow A$$

## Question 4

Using diagrams if you wish, write down the flight paths for

**1** 7 player, step 3          **3** 8 player, step 6

**2** 7 player, step 2

Try doing some more of your own until you can recognise a pattern in your results.

## Question 5

Without drawing diagrams, work out the flight path of the ball in each of these cases and work out which players receive the ball.

**1**  5 player,  step 4          **6** 23 player,  step 2

**2**  6 player,  step 2          **7** 11 player,  step 6

**3** 10 player,  step 5          **8**  9 player,  step 4

**4** 12 player,  step 4          **9** 14 player,  step 15

**5** 15 player,  step 6          **10** 14 player,  step 6

## Question 6

Combining your results from this investigation, can you offer a general method of determining the relationship between the number of players, the step size, the flight path and which players receive the ball.

### Extension work

In the main Passola work, the steps were always constant. Now investigate what happens when the steps vary. Here are a few examples, all illustrated with 8 player games.

## Example

(i) The step alternates between 1 and 2.

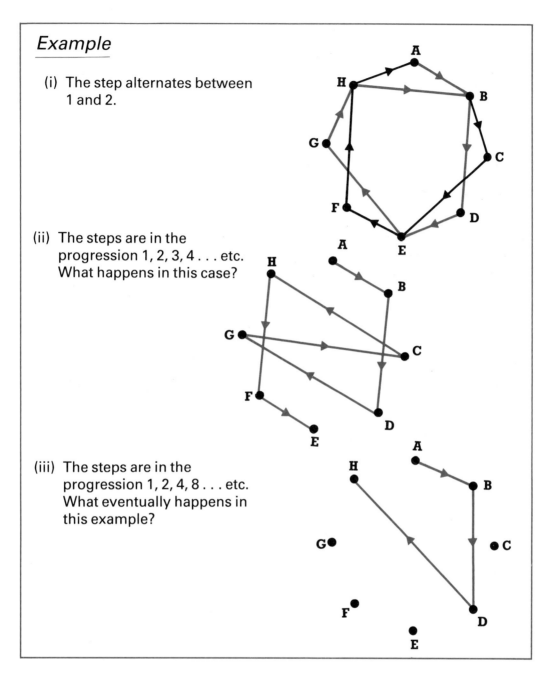

(ii) The steps are in the progression 1, 2, 3, 4 . . . etc. What happens in this case?

(iii) The steps are in the progression 1, 2, 4, 8 . . . etc. What eventually happens in this example?

You can invent any progression or pattern for the steps. Try a few for yourself. Report your findings for each one you try. In all cases try to predict:

● whether all players receive the ball,
● any ways of working out flight paths, receivers of the ball etc.,
● the conditions necessary for players to receive the ball.

# 12 Pythagoras 2

## Model building

You will need two 1 metre lengths of balsa or similar wooden struts, some adhesive and appropriate cutting instruments for this question.
The diagrams below show sketches of 3 shapes.

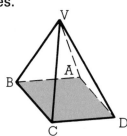

The vertex V is vertically above the mid-point of the horizontal square base ABCD.

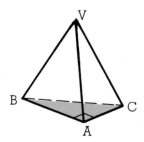

The horizontal base, ABC is a triangle with a right angle at A. The vertex V is vertically above the point A.

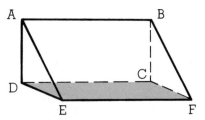

The horizontal base CDEF is a rectangle as is the vertical back ABCD.

Construct any 2 of the 3 models. In each of your models you must use all, or as much as possible, of the 1 metre balsa struts.

## Question 1

Classify the triangles, whose lengths of sides are given below, as being acute, right-angled or obtuse.

**1** (7, 8, 11)    **5** (15, 17, 18)

**2** (8, 6, 10)    **6** (48, 55, 73)

**3** (15, 8, 7)    **7** (9, 40, 41)

**4** (17, 19, 23)    **8** (8, 15, 19)

## Question 2

Calculate the distance between the following co-ordinates. Can you do it without drawing the points on a grid?

**1** (0,0) and (5,0)    **4** (0,0) and (4,3)

**2** (0,0) and (0,5)    **5** (3,4) and (5,7)

**3** (0,0) and (3,4)    **6** (−3,−2) and (2,10)

## Question 3

A perfect cube has a side of length 4 cm. Calculate the length of the diagonal from corner A to corner B.

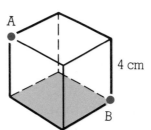

## Question 4

The dimensions of a cuboid are $6 \times 8 \times 11$ cm. Showing all of your working and any diagrams calculate the length of the longest diagonal of the cuboid.

## Question 5

The sketch shows a cone with circular base of radius $r = 16$ cm. The slant height $d$ of the cone is 34 cm. Find $h$. (Note that the vertex of the cone is vertically above the centre of the circular base.)

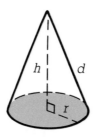

## Question 6

A cube has a volume of 27 cm³. Calculate the length of its longest diagonal.

## Question 7

The sketch shows a wedge ABCDEF. The horizontal base, CDEF, is a rectangle, as is the vertical back, ABCD. The angles ADE and BCF are both 90°. The lengths AD and BC are both 8 cm, whilst the lengths DE and CF are both 12 cm.

Calculate the lengths AE and BF. Given that BE = 18 cm, calculate the length of AB.

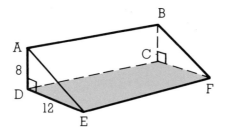

## Question 8

The sketch shows a tetrahedron VABC. The base is a triangle right-angled at A. The vertex V is vertically above A. Given that AB = 8 cm, VB = 17 cm and AC = 11 cm, calculate:

(i) the length AV
(ii) the length VC
(iii) the length BC.

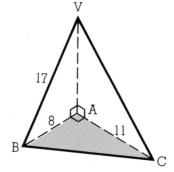

## Question 9

Imagine a square-based pyramid VABCD with vertex V vertically above the centre of the horizontal base ABCD. The distance AB = 8 cm and the distance VC = 17 cm.

Calculate:

(i) the distance BD
(ii) the vertical height of V above the horizontal base ABCD.

# 13 Frogs 1

You may have played the game of Frogs in SIGMA 1. Here is a reminder of the rules:

(i) a move can be either a slide or a jump. Move only 1 counter at a time

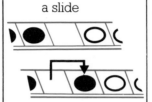
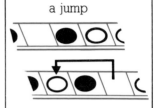

(ii) black only moves ➡
white only moves ⬅

(iii) a counter can slide into an empty adjacent space

(iv) a counter can jump over 1 — but no more — of the other colour to land in an empty space.

## Question 1

Using counters and a grid, start like this:

and try to play the game of Frogs so that you finish like this:

Count and record the number of moves it takes.

## Question 2

Repeat the game with 1, 2, 4, 5, 6 . . . counters of each colour. Count the moves and record your results in a table or using a method of you own.
Explain any strategies you use in playing the game.
Write down any observations you can make about the number patterns in your table of results.

## Question 3

How many moves would it take if you had

**1** 20 counters of each colour      **2** 50 counters of each colour?

## Question 4

Can you **generalise** the result for the relationship between the number of counters of each colour and the number of moves? The more ways you can express this relationship, the better.

## Question 5

Now try the game with *different* numbers of counters on each side.

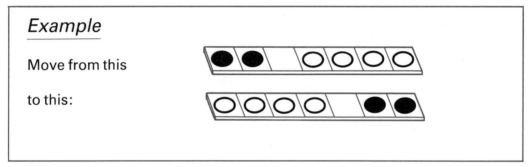

### Example

Move from this

to this:

Do as many examples as you want. Always count the numbers of moves it takes to complete the game. Record your results in a table and write down anything you notice about the results. Try to **generalise** the result for games with different numbers of counters on each side.

## Question 6

How many moves would it take if you had 20 counters of one colour and 30 of another on opposite sides?

## Centre-point

Rules:

(i) move only one counter at a time

(ii) counters can move in any direction

(iii) a counter can slide into an empty adjacent space

(iv) a counter can jump over 1 — but no more — of the other counters to land in an empty space

(v) any counter that is jumped over must be removed.

# Question 7

Start like this:

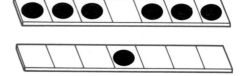

and try to finish, in as few moves as
possible, like this:

Record all your results in whatever way you think best. You must always
start with an even number of counters, and have an equal number either side
of the centre point.

Try the game with different numbers of counters.

# Question 8

Find the relationship between the number of counters at the start and the
least number of moves required to complete the game. You will need to play
the game several times with different numbers of counters and to record all
your results. Look at them carefully and try to find a **general relationship**.

# 14 Symmetry

## Reflections and rotations

### Question 1

Copy these shapes on to dotted paper and fill in the reflection in the dotted mirror line.

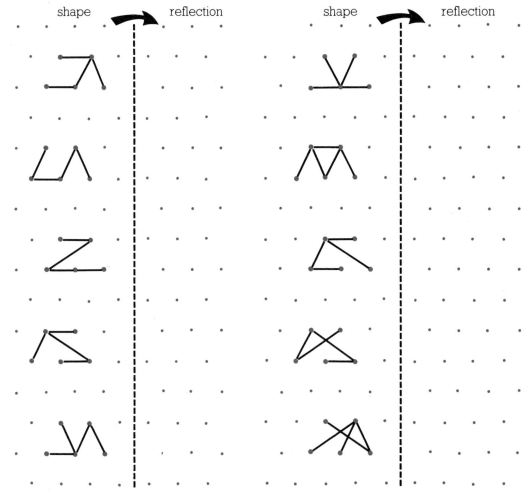

shape → reflection        shape → reflection

# Question 2

After copying these shapes on to dotted paper, fill in the reflection.

shape        reflection        shape        reflection

## Question 3

Rotate the lines or shapes through 180° about the centre point shown:

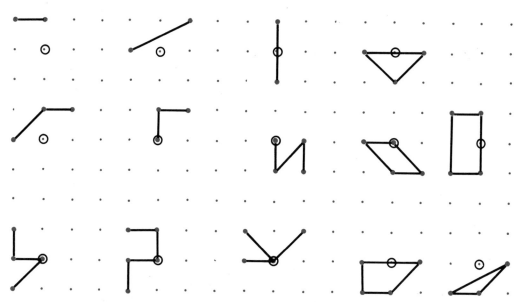

## Question 4

Rotate these shapes through 90° turns about the circled point shown.

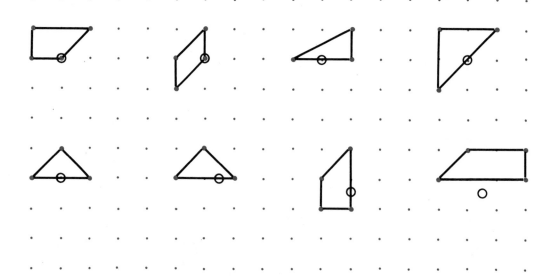

# Question 5

Rotate these shapes through 270° about the circled point shown.

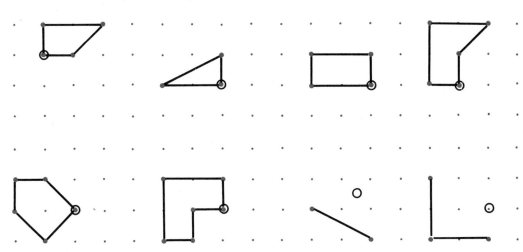

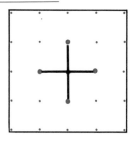

# Question 6

Starting like this: the 5 × 5 board can be cut up into 4 (identical) **congruent** pieces. It has **rotational symmetry**.

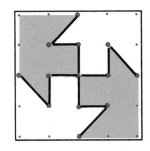

Now think of your own way of dividing the 5 × 5 board into 4 congruent pieces so that there is rotational symmetry. There are 13 different ways starting with the cross above and 10 different ways starting with the cross on the right.

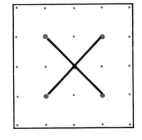

# Question 7

Here is a way of cutting up the board into 4 congruent pieces so that there is **reflection symmetry**. Can you find 7 more ways?

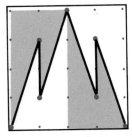

# 15 Diagonals

In this Chapter, your task is to investigate the relationship between the number of vertices and the number of diagonals of a regular polygon.

A **diagonal** is a line joining 2 **vertices** which is not one of the sides of the polygon. For instance, in a regular hexagon (6 vertices), 3 diagonals are shown.

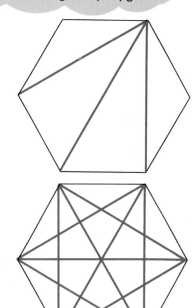

If we draw in all the diagonals we have a total of 9.

This gives the mapping:

| number of vertices | number of diagonals |
|---|---|
| 6 ⟶ | 9 |

## Question 1

Look at the number of diagonals on each of the following regular polygons.

**1** a square      (4 vertices)      **3** a heptagon      (7 vertices)

**2** a pentagon      (5 vertices)      **4** an octagon      (8 vertices)

Think of some other polygons as well and set out your results in a table like the one shown.

| number of vertices | number of diagonals |
|---|---|
| 3 | |
| 4 | |
| 5 | |
| 6 ———————————————→ | 9 |
| 7 | |
| 8 | |
| ⋮ | |

Write down any patterns you see in this table of results. Can you **generalise** the result so that you can predict the number of diagonals in any size polygon? Can you offer an explanation or proof of the generalisation?

## Question 2

Calculate the number of diagonals in a regular polygon having

**1** 10 vertices          **3** 50 vertices

**2** 20 vertices          **4** 1000 vertices

## Question 3

Calculate the number of vertices of the regular polygon given that the number of diagonals is:

**1** 65        **2** 90        **3** 495        **4** 5150

## Question 4

There is a regular polygon for which:
$$\begin{bmatrix} \text{number of} \\ \text{vertices} \end{bmatrix} = \begin{bmatrix} \text{number of} \\ \text{diagonals} \end{bmatrix}$$

(i) State the number of vertices of that polygon.

(ii) Prove that there is only one such polygon.

## Question 5

Extend this problem to investigate the number of 3 dimensional diagonals in a regular solid. You may want to make the solids out of Constructa-straws, balsa struts or other similar materials.

# 16 Cuisenaire Activity 3

w : white
r : red
g : light green
p : pink
y : yellow
d : dark green
b : black
t : tan (brown)
B : blue
O : orange

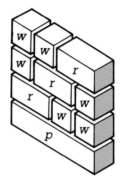

## Partitions

The diagram shows the ways that you can partition the *light green* rod:

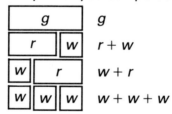

g

r + w

w + r

w + w + w

## Question 1

(i) How many different ways can you find to partition the *pink* rod? How can you be sure you've got them all?

(ii) How many partitions of the *yellow* rod are there? Use a system to list all the possibilities.

For the *light green* partitions (shown above) there is:

*1* way using 1 rod
*2* ways using 2 rods
*1* way using 3 rods

Look at the pattern of numbers like the *1, 2, 1* shown here, for answers to (i) and (ii). From these can you predict the pattern of numbers for the partitions of the *dark green* rod?

## Question 2

You have to make each rod using 3 other rods in as many different ways as you can.

*Light green* can be made only 1 way

*Pink* can be made in 3 different ways

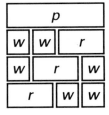

Can you make *yellow* in 6 different ways? Remember that you can only use 3 rods.

## Question 3

Using 3 rods in how many different ways can you make:

(i)  the *dark green* rod
(ii)  the *black* rod and so on . . .?

Can you see a pattern in these numbers? Can you see how you get this pattern of numbers from the way the rods are built up?

## Question 4

For the *pink* rod

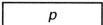

there are five different ways of using *reds* and *whites* to make the *pink*.

| | r | | | r | |
|---|---|---|---|---|---|
| | | w | w | r | |
| | w | r | | w | |
| | r | | w | w | |
| | w | w | w | w | |

How many different ways are there, using only *reds* and *whites*, to make:

| | | | | | | |
|---|---|---|---|---|---|---|
| **1** | *white* | 1 | | **6** | *dark green* | ? |
| **2** | *red* | ? | | **7** | *black* | ? |
| **3** | *light green* | ? | | **8** | *tan* | ? |
| **4** | *pink* | 5 | | **9** | *blue* | ? |
| **5** | *yellow* | ? | | **10** | *orange* | ? |

## Question 5

For the *pink*, the number of ways using 2 *reds* = *1*
using 1 *red* = *3*
using 0 *reds* = *1*

Look at the number patterns, like the *1, 3, 1* shown here, for your answers to question 4.

## Question 6

Repeat what you have done using *light green*s and *white*s instead of *red*s and *white*s. Compare your answers to those from question 4 and question 5.

## Cover-up

## Question 7

2 players are needed for this game. Draw a 5 × 5 grid and take it in turns to place *red* rods on the grid so that each rod covers 2 squares. A player wins when the other player cannot go.

What is the least number of goes in which you can win? What is the most? Can you find a way of winning the game if the other player goes first?

## Question 8

Try the same game on a different grid:

What is the least possible number of goes?

Look at the least possible number of goes on other grids of different sizes.

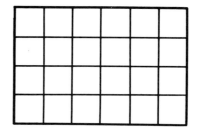

## Question 9

Play the game using *light green* rods.

## Question 10

Invent a game of your own played on a grid using the Cuisenaire rods.

## Question 11

In the diagram of a 5 × 5 grid, *red* rods have been arranged to leave the maximum number of spaces empty with *no 2 spaces beside one another*. The spaces are surrounded either by rods or the edge of the grid. As you can see, there are 7 spaces.

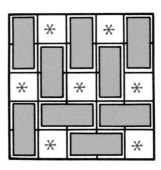

Arrange *red* rods on grids of sizes 3 × 3, 4 × 4, 6 × 6 and 7 × 7 to leave the maximum number of spaces as in the 5 × 5 grid.

Try some larger grids as well. Is there a relationship between the maximum number of spaces and the grid size?

Do the same using *light green* rods instead of *red* ones. How do the two sets of results compare?

# 17 Quadratic Equations

We saw in chapter 46 (Quadratics—Zeros) in SIGMA 1 how a quadratic equation might arise from some investigation such as Frogs. To remind you of the rules in Frogs look at chapter 13. If we have $x$ white and $x$ black counters, then the total number of moves is:

$$x(x + 2) \quad \text{or} \quad x^2 + 2x$$

Suppose we say that with equal numbers of white and black counters the game was done in 24 moves – and then pose the question:

'How many counters did we have of each colour?'

we are in fact asking you to solve the equation

$$x^2 + 2x = 24$$

or

$$x^2 + 2x - 24 = 0$$

This is called a **quadratic equation**.

If you repeat the work of SIGMA 1 you will find that the zeros of this quadratic are:

$$4 \text{ and } -6$$

If we put $x = 4$ into the equation we get

$$4^2 + 2(4) - 24$$
$$16 + 8 - 24$$
$$24 - 24 = 0$$

Putting $x = -6$

$$(-6)^2 + 2(-6) - 24$$
$$36 + (-12) - 24$$
$$36 - 12 - 24$$
$$24 - 24 = 0$$

So putting $x = 4$ or $x = -6$ into the equation makes it equal 0.

The zeros are usually called the **solutions** of the quadratic equation. All quadratic equations have 2 solutions.

We are now going to look at some ways of finding the solutions of quadratic equations.

## Spotting the solution

In some cases, when the equation is simple enough, we can spot the solution.

> ### *Example*
>
> (i) $x^2 - 25 = 0$
> $$x^2 = 25$$
> $$x = \sqrt{25}$$
> so $x = 5$ or $-5$
>
> (ii) $(x + 3)^2 = 49$
> $$x + 3 = \sqrt{49}$$
> $$x + 3 = 7 \text{ or } -7$$
>
> so
>
> $x + 3 = 7$
> $x = 7 - 3$
> $x = 4$
>
> or
>
> $x + 3 = -7$
> $x = -7 - 3$
> $x = -10$
>
> (iii) $x^2 - 6x = 0$
> $$x^2 = 6x$$
>
> You ought to spot that we can cancel an $x$ to give $x = 6$ as a solution. The second solution is $x = 0$ because
> $$0^2 = 6 \times 0 = 0$$
> so the solutions are $x = 6$ and $x = 0$

## Question 1

Solve each of the following equations.

**1** $x^2 - 16 = 0$ **2** $49 - x^2 = 0$

**3** $(3x + 2)^2 = 25$

**4** $(x + 5)^2 = 4$

**5** $(2x + 3)^2 = 36$

**6** $x^2 - 7x = 0$

**7** $2x - x^2 = 0$

**8** $9 - x^2 = 0$

**9** $(x - 4)^2 = 64$

**10** $(3x - 2)^2 = 16$

Of course you might be able to spot the solution of any quadratic, but it is unlikely. However, if you can recall the work on the zeros of a quadratic which you did in SIGMA 1 it will be helpful. To remind you, if the quadratic is

$$x^2 - 8x + 15 = 0$$

then the solutions are $x = 3$ and $x = 5$. Note:

$$-8 = -(3 + 5)$$
$$+15 = (3 \times 5)$$

Not a coincidence!

Look again at $x^2 + 2x - 24 = 0$

$$+2 = -(4 + (-6))$$
$$-24 = (4 \times (-6))$$

So

$$x = 4 \text{ and } x = -6$$

The result is that if the quadratic equation

$$x^2 + ax + b = 0$$

has solutions $x = n$ and $x = m$, then

$$a = -(n + m) \text{ and } b = n \times m$$

## Question 2

Try to use this result to spot the solutions to these equations.

**1** $x^2 - 7x + 12 = 0$

**2** $x^2 - 10x + 21 = 0$

**3** $x^2 + 8x + 15 = 0$

**4** $x^2 + 8x + 12 = 0$

**5** $x^2 - 6x + 8 = 0$

**6** $x^2 + 2x - 15 = 0$

**7** $x^2 + 13x + 42 = 0$

**8** $x^2 - 3x - 28 = 0$

**9** $x^2 - 10x + 24 = 0$

**10** $x^2 - 2x - 15 = 0$

## Completing the square

When we cannot spot the solutions we need a technique for solving the equation. There are many — but without doubt the most powerful technique is by completing the square. The idea with this method is to convert an equation into the easily-solved form

$$z^2 = \text{a constant.}$$

When the equation is in this form, say $z^2 = 10$, the equation is easily solved:

$$z = + \sqrt{10} \text{ and } z = - \sqrt{10}$$

Few quadratic equations are in this form so adjustments have to be made to them. For example,

$$x^2 + 2x - 3 = 0$$

can be transformed by completing the square, into:

$$(x + 1)^2 = 4$$

In this form, the solutions are easily calculated.

$$x + 1 = + \sqrt{4} = 2$$

so $\qquad\qquad x = 1$

and $\qquad\qquad x + 1 = - \sqrt{4} = -2$

so $\qquad\qquad x = -3$

Now let's look at the process of completing the square. Suppose we have the expression:

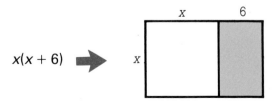

This is the same as:

$x(x + 6)$

By halving the width of the second rectangle and moving one of the end strips we have:

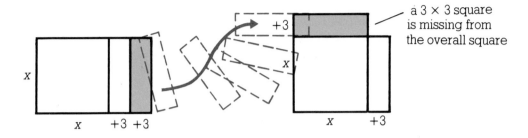

a 3 × 3 square is missing from the overall square

So the area of this shape is:

$$(x + 3)^2 - 3^2 \quad \text{or} \quad (x + 3)^2 - 9$$

So, before the change we had a rectangle of area:

$$x(x + 6) \text{ or } x^2 + 6x$$

and afterwards we had a shape of area

$$(x + 3)^2 - 9$$

but these areas are the same, so we can say

$$x^2 + 6x = (x + 3)^2 - 9$$

We have completed the square on $x^2 + 6x$
Similarly for $x^2 + 8x$:

$$x^2 + 8x = x(x + 8)$$

This gives the following rectangle:

a 4 × 4 square is missing

area = $x(x+8)$

The missing square is 4 × 4

So

$$x^2 + 8x = (x + 4)^2 - 16$$

In general terms, to complete the square on:

$$x^2 + kx$$

we write:

$$\left(x + \frac{k}{2}\right)^2 - \left(\frac{k}{2}\right)^2$$

Using this method

$$x^2 + 12x = (x + 6)^2 - 36$$

$$x^2 + 10x = (x + 5)^2 - 25$$

$$x^2 - 10x = (x + (-5))^2 - (-5)^2$$

$$= (x - 5)^2 - 25$$

Now we can apply this method to solving quadratic equations.

## Example

(i) To solve:

$$x^2 + 6x - 16 = 0$$

we say:

$$x^2 + 6x = 16$$

Now we complete the square on $x^2 + 6x$:

$$x^2 + 6x = (x + 3)^2 - 9$$

So,

$$x^2 + 6x = 16$$

becomes

$$(x + 3)^2 - 9 = 16$$

$$(x + 3)^2 = 25$$

so

$$x + 3 = 5$$

or

$$x + 3 = -5$$

therefore

$$x = 2 \text{ or } x = -8$$

(ii) $$2x^2 - 7x + 3 = 0$$

Divide by the 2 to make the coefficient of $x^2 = 1$.

$$\frac{2x^2}{2} - \frac{7x}{2} + \frac{3}{2} = \frac{0}{2}$$

$$x^2 - \frac{7x}{2} + \frac{3}{2} = 0$$

$$x^2 - \frac{7x}{2} = -\frac{3}{2}$$

Completing the square on $x^2 - \frac{7x}{2}$ gives:

$$\left(x - \frac{7}{4}\right)^2 - \frac{49}{16}$$

$$\left(x - \frac{7}{4}\right)^2 - \frac{49}{16} = -\frac{3}{2}$$

$$\left(x - \frac{7}{4}\right)^2 = -\frac{3}{2} + \frac{49}{16}$$

$$\left(x - \frac{7}{4}\right)^2 = \frac{25}{16}$$

$$x - \frac{7}{4} = \frac{5}{4} \text{ or } -\frac{5}{4}$$

$$x = \frac{5}{4} + \frac{7}{4} = \frac{12}{4} \quad \text{ or } x = -\frac{5}{4} + \frac{7}{4} = \frac{2}{4}$$

So the solutions are $x = 3$ or $x = \frac{1}{2}$

## Question 3

Solve each of the following equations by completing the square

1  $x^2 - 8x + 12 = 0$

2  $x^2 + 10x + 21 = 0$

3  $x^2 - 6x - 16 = 0$

4  $x^2 + 14x + 48 = 0$

5  $x^2 - 7x + 10 = 0$

6  $24 - 11x + x^2 = 0$

7  $x^2 - 8x - 12 = 0$

8  $x^2 + x - 17 = 0$

9  $2x^2 - 13x + 15 = 0$

10  $3x^2 - 11x + 10 = 0$

## Example

(i) Suppose we wish to solve

$$x^2 + 2x - 24 = 0$$

We can say

$$x^2 + 2x = 24$$

so

$$x(x + 2) = 24$$

If we make

$$y = x + 1$$

then

$$x + 2 = y + 1$$

and

$$x = y - 1$$

This is the **transformation**. The question then becomes

$$(y - 1)(y + 1) = 24$$

This is the difference between 2 squares (see Chapter 21)

$$y^2 - 1 = 24$$
$$y^2 = 25$$
$$y = 5 \text{ or } -5$$

so

$$x + 1 = 5 \text{ or } x + 1 = -5$$

Therefore

$$x = 4 \text{ or } x = -6$$

(ii) Here is another example:

$$x^2 - 8x = 15 = 0$$
$$x^2 - 8x = -15$$
$$x(x - 8) = -15$$

Now we do the transformation:

$$y = x - 4$$

This gives:

$$(y + 4)(y - 4) = -15$$
$$y^2 - 16 = -15$$

Again this gives the difference between 2 squares. Work through this example and find the solutions of $x = 5$ and $x = 3$.

(iii) This third example is much harder.

$$3x^2 - 16x + 5 = 0$$
$$3x^2 - 16x = -5$$
$$x(3x - 16) = -5$$

Multiply by 3 to get

$$3x(3x - 16) = -5 \times 3$$
$$3x(3x - 16) = -15$$

For the transformation put $y = 3x - 8$ so

$$3x = y + 8$$

and

$$3x - 16 = y - 8$$

Therefore

$$(y + 8)(y - 8) = -15$$
$$y^2 - 64 = -15$$
$$y^2 = -15 + 64 = 49$$

So

| | | |
|---|---|---|
| $y = 7$ | or | $y = -7$ |
| $3x - 8 = 7$ | | $3x - 8 = -7$ |
| $3x = 7 + 8$ | | $3x = -7 + 8$ |
| $3x = 15$ | | $3x = 1$ |
| $x = \dfrac{15}{3}$ | | $x = \dfrac{1}{3}$ |
| $x = 5$ | | $x = \dfrac{1}{3}$ |

The problem with this method, is working out what the transformation is. Look at this list and see if you can spot how the transformation is worked out.

| re-arranged equation | transformation |
| --- | --- |
| $x(x + 2) = 24$ | $y = x + 1$ |
| $x(x - 8) = -15$ | $y = x - 4$ |
| $3x(3x - 16) = -15$ | $y = 3x - 8$ |

# Question 4

Solve these quadratic equations by transformations.

1  $x^2 - 7x + 12 = 0$

2  $x^2 + 6x + 8 = 0$

3  $x^2 - 6x - 16 = 0$

4  $2x^2 - 7x + 3 = 0$

5  $3x^2 - 11x + 10 = 0$

# Question 5

You are now given 10 quadratics which you can solve in any way you wish.
Try the same question a few times using different methods.

1  $x^2 - x - 2 = 0$

2  $x^2 + x - 12 = 0$

3  $x^2 - 3x - 10 = 0$

4  $x^2 - 3x - 10 = 0$

5  $x^2 + 4x - 32 = 0$

6  $2x^2 + 11x + 5 = 0$

7  $9 + x - x^2 = 0$

8  $15 + 2x^2 - 11x = 0$

9  $x^2 - 8x + 16 = 0$

10  $8x^2 - 26x + 21 = 0$

# Question 6

What happens when you try to solve the equation:

$$x^2 - 8x + 41 = 0 \quad ?$$

# 18 Graphical Transform 2

## Sketching Quadratics

*Example:* Sketch $y = (x + 3)^2 - 2$ using 2 or 3 stages:

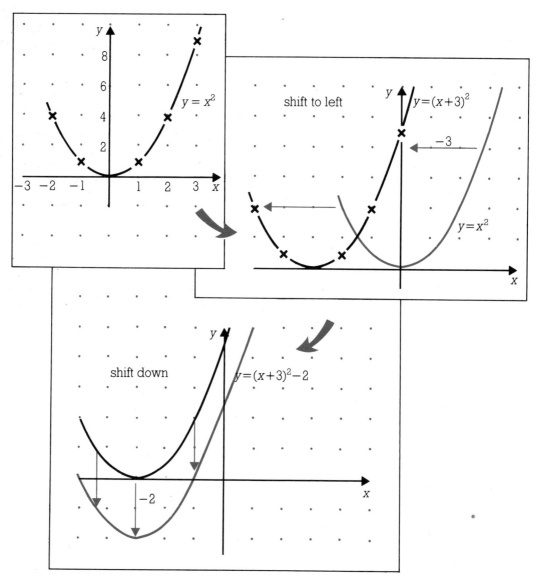

# Question 1

Use 2 or 3 sketches starting from $y = x^2$ in each case, to draw the following lines:

**1** $y = x^2 + 5$

**2** $y = x^2 - 2$

**3** $y = x^2 + \frac{1}{2}$

**4** $y = (x+3)^2$

**5** $y = (x-2)^2$

**6** $y = (x+20)^2$

**7** $y = (x+2)^2 + 3$

**8** $y = (x+5)^2 - 2$

**9** $y = (x-6)^2 + 9$

**10** $y = (x-\frac{3}{4})^2 - \frac{2}{3}$

In the following diagrams, notice how the shape of the graphs change with small differences in the equation.

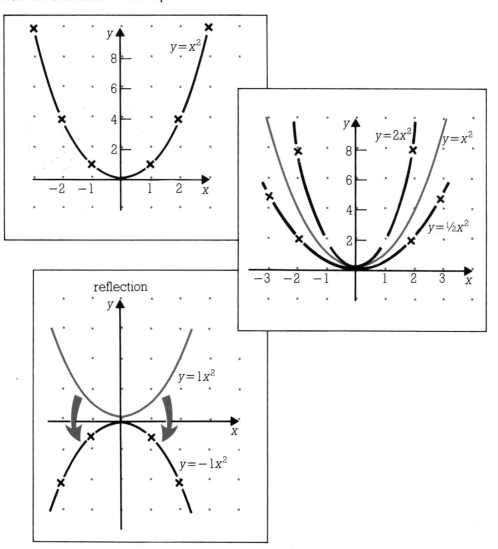

## Question 2

Use 2 or 3 sketches to draw the following curves.

1  $y = 3x^2$                 7  $y = \frac{1}{2}x^2 - 3$

2  $y = 10x^2$               8  $y = 2(x+1)^2$

3  $y = \frac{1}{2}x^2$         9  $y = 8(x-3)^2$

4  $y = -2x^2$             10  $y = -(x+5)^2$

5  $y = 2x^2 + 1$         11  $y = 7(x+9)^2 - 16$

6  $y = 10x^2 + 5$       12  $y = \frac{3}{4}(x-\frac{4}{3})^2 + \frac{2}{3}$

## Question 3

Take care when sketching these graphs. They are much more difficult than the previous questions. First check that the following sketch is correct.

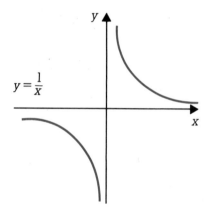

$y = \frac{1}{x}$

1  $y = \dfrac{1}{x + 2}$             6  $y = \dfrac{8}{x - 3}$

2  $y = \dfrac{1}{x + 9}$             7  $y = \dfrac{-5}{x + 6}$

3  $y = \dfrac{1}{x - 3}$             8  $y = \dfrac{x + 9}{x + 5}$

4  $y = 5\left(\dfrac{1}{x}\right)$        9  $y = \dfrac{x - 3}{x + 8}$

5  $y = \dfrac{-1}{x}$                10  $y = \dfrac{x - 6}{x - 4}$

# 19 Groups 1

The questions in this Chapter are about 4 different **systems**. When you have done them, try to see what they have in common. What do you think a system is?

## Odds and evens

$O$ is an odd number
$E$ is an even number

## Question 1

$O + O = ?$
$E + O = ?$

complete the table:

| + | O | E |
|---|---|---|
| O |   |   |
| E |   |   |

$O \times O = ?$
$E \times O = ?$

complete the table:

| × | O | E |
|---|---|---|
| O |   |   |
| E |   |   |

## Question 2

Use your tables to answer these:

**1** $O \times O \times O = ?$

**2** $O \times (O + E) = ?$

**3** $(O \times O) + (O \times E) = ?$

**4** $(E + O) \times (E + O) = ?$

**5** $O \times O \times O \times E = ?$

**6** $E^3 = E \times E \times E = ?$

**7** $O^2E^2 = O \times O \times E \times E = ?$

**8** $E + (O \times (E + O)) = ?$

**9** $(O + E)(O + E) = ?$

**10** $O^3E = O \times O \times O \times E = ?$

**11** $O \times O \times (O + ?) = E$

**12** $(E + E) \times ? = E$

**13** $O \times O \times O \times (E + ?) = O$

**14** $(O + ? + O) \times O = O$

**15** $(O + E + E + O + ?) \times O = O$

**16** $O \times (O + E)^2 = ?$

**17** $(O + E) \times (O + ?) = O$

**18** $O + (E \times (O + ?)) = O$

**19** $O^5 = ?$

**20** $8(O) = ?$

## Coins

Place 3 coins on a table, with the centre one showing a tail (T), and the others showing heads (H):

You can either turn no coins or turn any 2 coins at a time. There are 4 moves shown below.

move **A**: H T H → H T H

move **B**: H T H → T H H

move **C**: H T H → H H T

move **D**: H T H → T T T

---

*Example*

**C • D**

This means do move **C** followed by move **D**. In move **C**, the first coin stays the same and the others change. So starting with H T H we have:

H T H → H H T

In move **D**, the first and third change, so we have:

H H T → T H H

In going from H T H to T H H, we have changed the first 2 coins. This is the same as move **B**. Therefore:

**C • D = B**

---

# Question 3

Complete the following table:

| • | A | B | C | D |
|---|---|---|---|---|
| A | | | | |
| B | | | | |
| C | | | | |
| D | | | | |

## Change form

Look at these 4 ways of changing the arrangement of the shapes. There can be a change in colour **C** (pink or white), in shape **S** (circle or triangle) or both colour and shape **SC**.

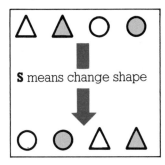

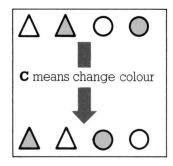

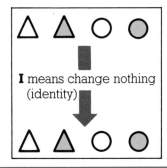

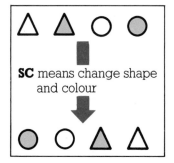

---

*Example*

$$S \bullet C = SC$$
$$SC \bullet C = S$$
$$C \bullet C = I$$

## Question 4

Complete the combination table:

| • | S | C | SC | I |
|----|----|----|----|----|
| S | | SC | | |
| C | | I | | |
| SC | | S | | |
| I | | | | |

## Question 5

Use the combination table to work out:

1  $S \cdot ? = C$

2  $C \cdot ? = SC$

3  $C \cdot (S \cdot C) = ?$

4  $S \cdot (SC \cdot SC) = ?$

5  $C \cdot (C \cdot ?) = S$

6  $S \cdot (C \cdot S) = ?$

7  $(S \cdot C) \cdot S = ?$

8  $S \cdot (SC \cdot ?) = C$

9  $C \cdot (C \cdot ?) = SC$

10  $(SC \cdot ?) \cdot S = C$

11  $S^5 = ?$

12  $SC^3 = ?$

## Question 6

The tables you have completed in the previous questions are called operation tables. Complete the following operation tables for each of the 5 systems shown.

(i) Cyclic permutations

$x$:  $(a, b, c) \rightarrow (a, b, c)$
$y$:  $(a, b, c) \rightarrow (b, c, a)$
$z$:  $(a, b, c) \rightarrow (c, a, b)$

| • | x | y | z |
|----|----|----|----|
| x | | | |
| y | | | |
| z | | | |

(ii) Addition, Mod 3

| + | 0 | 1 | 2 |
|---|---|---|---|
| 0 | | | |
| 1 | | | |
| 2 | | | |

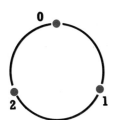

(iii) Multiplication, Mod 7

| × | 1 | 2 | 4 |
|---|---|---|---|
| 1 | | | |
| 2 | | | |
| 4 | | | |

(iv) Rotations of an equilateral triangle.

$R_0$:  rotation of $0°$
$R_{120}$: rotation of $120°$
$R_{240}$: rotation of $240°$

| • | $R_0$ | $R_{120}$ | $R_{240}$ |
|---|---|---|---|
| $R_0$ | | | |
| $R_{120}$ | | | |
| $R_{240}$ | | | |

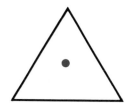

What do you notice about these tables?

# 20 Combined Reflections

## Question 1

In this exercise reflect the shape in the $x = 3$ line and then reflect this shape in the $x = 7$ line. Copy the grids into your books and draw the 2 reflections for each question.

**1**

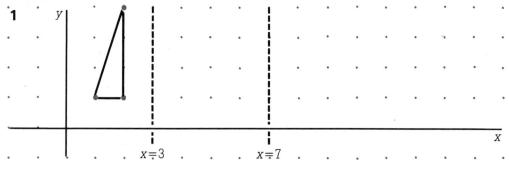

**2**

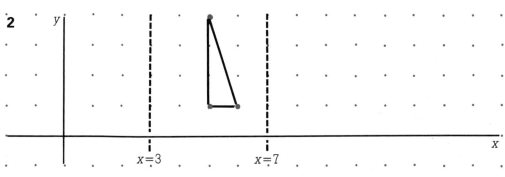

**3**

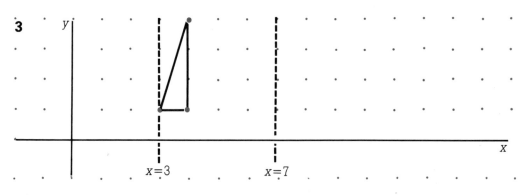

How far does the triangle get shifted after the 2 reflections? What has this to do with the distance between mirror lines?

## Question 2

Reflect the shape in the $y = 4$ line then reflect this shape in the $y = 9$ line. Copy the grids into your books and draw the 2 reflections.

How far has the shape been shifted after the two reflections?

## Question 3

In **1** and **2** reflect the shape first in the $y = x$ line and then reflect this shape in the $x$-axis. Draw the results on dotted paper or graph paper.

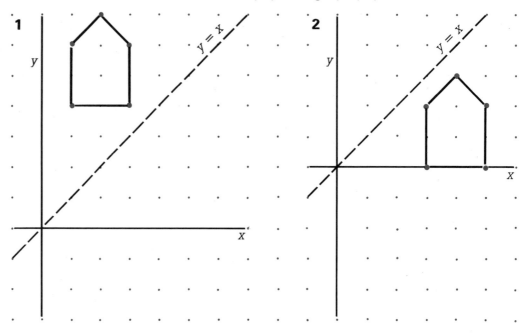

Reflect in $y = x$ first then reflect in the $y$-axis.

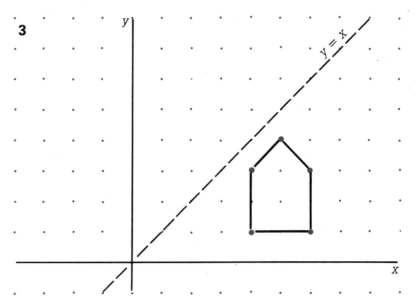

What rotation has the same effect as the combined reflections?

In the diagram you can reflect in:
    line $M_1$
    line $M_2$
or line $M_3$

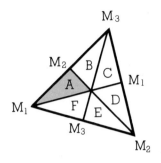

| *Example* |
| :--- |
| Reflections can be represented like this. |
| $M_2 (A) = B$ |

## Question 4

Do the questions below remembering that where there is more than 1 reflection to do (e.g. $M_2 M_3 (A)$), you do the reflection nearest the bracket first (in this case, $M_3$).

**1** $M_3 (A)$

**2** $M_1 (A)$

**3** $M_2 (B)$

**4** $M_2 (C)$

**5** $M_3 (F)$

**6** $M_3 (D)$

**7** $M_2 (D)$

**8** $M_1 (B)$

**9** $M_1 (C)$

**10** $M_2 (E)$

**11** $M_2 M_3 (A)$

**12** $M_3 M_2 (A)$

**13** $M_1 M_2 (A)$

**14** $M_2 M_1 (A)$

**15** $M_2 M_2 (A)$

**16** $M_3 M_1 M_3 (A)$

**17** $M_1 M_2 M_3 (A)$

**18** $M_2 M_3 M_1 (A)$

**19** $M_1 M_2 M_3 M_1 (A)$

**20** $M_1 M_2 M_3 M_2 M_1 (A)$

Which positions can you get to from A using an even number of reflections and which need an odd number?

# 21 The Difference Between 2 Squares

An example of a **square number** is 25 because it is

$$5 \times 5 = 5^2,$$

and can be set out in a square as dots.

Similarly, 9 is also a *square number*, because it is $3 \times 3 = 3^2$, and can be set out as:

25 − 9 can be written as $5^2 - 3^2$ and can be set out as:

You could have taken away *any* 9 dots. It leaves, in this case, an L shape.

By rearranging the remaining 16 dots we can construct a rectangle of 16 dots in an 8 × 2 configuration.

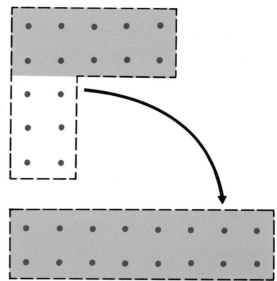

The length of this rectangle is 8 dots and its width is 2 dots. So the difference between the 2 square numbers 25 and 9 is the rectangular number 16.

Your task is to look at some other *differences between 2 squares* and to find a **generalised** result.

88

First, let's re-cap our result.

$$25 - 9 = 16$$
$$25 - 9 = 8 \times 2$$
$$5^2 - 3^2 = 8 \times 2$$

You might, even at this stage, spot a relationship between the numbers.

# Question 1

Try each of the following, using counters, dot patterns or a method of your own.

| | | | | | | |
|---|---|---|---|---|---|---|
| **1** $25 - 4$ | or | $5^2 - 2^2$ | **4** $100 - 36$ | or | $10^2 - 6^2$ |
| **2** $36 - 9$ | or | $6^2 - 3^2$ | **5** $16 - 1$ | or | $4^2 - 1^2$ |
| **3** $49 - 25$ | or | $7^2 - 5^2$ | | | |

Make up 3 others of your own.

- Can you *spot a pattern* in your results?
- Can you **generalise** the result?
- Can you write your generalisation in symbolic form?
- Can you give a proof of the generalisation?

# Question 2

Use the result from question 1 to calculate:

**1** $17^2 - 7^2$

**2** $64^2 - 36^2$

**3** $(6.57)^2 - (3.43)^2$

**4** $(9.66)^2 - (0.34)^2$

**5** $(12.74)^2 - (7.26)^2$

**6** $(11.65)^2 - (7.35)^2$

# Question 3

*Factorise* the following expressions:

**1** $x^2 - y^2$

**2** $p^2 - q^2$

**3** $9x^2 - 4y^2$

**4** $n^2 - 25$

**5** $36 - 4x^2$

**6** $16a^2 - b^2$

**7** $49t^2 - 25x^2$

**8** $4 - 25y^2$

## Extension problems

The generalised result for the difference of 2 squares is a useful aid to solving these problems. You may however, want to solve the problems by your own methods.

These extension problems are not easy, so feel free to work in groups and discuss any problems you encounter.

## Question 4

The sum of 2 numbers is 12 and the difference between them is 3. Calculate the difference between the squares of the 2 numbers.

## Question 5

Suppose that

$$4x^2 - y^2 = 32$$

and

$$2x - y = 4$$

Calculate the values of $x$ and $y$.

## Question 6

Solve the simultaneous equations:

$$25x^2 - y^2 = 16$$

and

$$5x + y = 8$$

## Question 7

You have found a generalised result for the difference of 2 squares. Can you now find a generalised result for the difference of cubes and express it in symbolic form starting with

$$n^3 - m^3 \ ?$$

# 22 Road Inspection

In this chapter, we will deal with the problems of inspecting roads, railway tracks, pipes or snow gritting. In all these activities, it is important to move along each stretch of road, track or pipe at least once, always returning to the starting point. The problem is to do this with as little backtracking and repeating of routes as possible.

---

## Example

Consider a snowgritting lorry which has to grit all the roads shown on the diagram below. As these are small roads, the gritter only has to grit each road once (covering both sides of the road as it does it). Can you design a route, starting and finishing at the depot, which covers every road only once? All the distances in this and other diagrams are in kilometres.

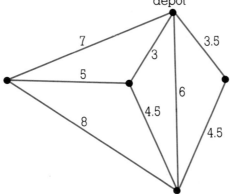

It shouldn't take you long to realise that it can't be done. At least one road will have to be repeated.

---

## Question 1

Design a route, starting and finishing at the depot, which passes along each road at least once. There are several ways in which we can design a suitable route, but the *total* distance travelled will vary. For example, two possible routes are shown on the next page:

91

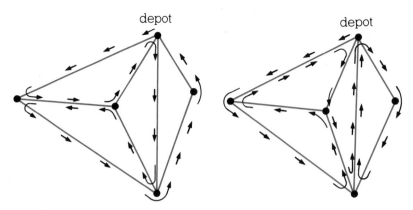

Clearly the first route is the best to take. It only repeats 1 of the roads.

Let's look at a straightforward problem. Suppose we wish to start and finish at Exeter and move along each road shown in the network below, but want the total distance travelled to be as small as possible.

If we check the number of roads at each vertex, we get:

| | |
|---|---|
| Okehampton | 2 |
| Exeter | 4 |
| Tavistock | 4 |
| Ashburton | 4 |
| Plymouth | 3 |
| Dartmouth | 3 |

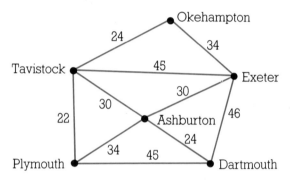

Only Plymough and Dartmouth have an *odd* number of roads. So we must repeat 1 road into Plymouth and 1 road into Dartmouth. So we duplicate the road Plymouth–Dartmouth, giving the route shown below.

The total distance travelled is the total of all the road lengths plus 45 kilometres (the repeated Plymouth–Dartmouth distance), giving a total of 379 kilometres.

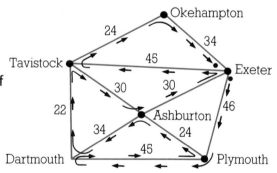

The problem becomes more difficult if we have more than 2 odd vertices. For example, if we also include an Okehampton–Ashburton road, as in the diagram at the top of the next page, we have 4 towns with odd numbers of roads; Okehampton, Ashburton, Plymouth and Dartmouth.

| Okehampton | 3 |
|---|---|
| Exeter | 4 |
| Tavistock | 4 |
| Ashburton | 5 |
| Plymouth | 3 |
| Dartmouth | 3 |

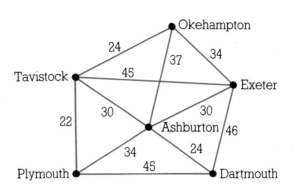

To complete a route plan which moves along every road at least once, we must consider an extra route between these odd towns, joining them up in pairs. This gives 3 possibilities:

| towns joined | extra distance travelled |
|---|---|
| (i) Okehampton–Ashburton and Plymouth–Dartmouth | 37 + 45 = 82 |
| (ii) Okehampton–Plymouth and Ashburton–Dartmouth | (24 + 22) + 24 = 70 |
| (iii) Okehampton–Dartmouth and Ashburton–Plymouth | (37 + 24) + 34 = 95 |

So the best plan is (ii) and we must repeat Okehampton–Plymouth (via Tavistock) and Ashburton–Dartmouth. This gives the following route plan:

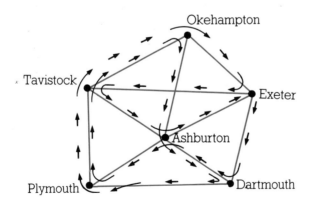

Total distance for the plan = 441 kilometres.

Now that we have a strategy for solving this type of problem, try the following exercises.

# Question 2

Find a route inspection plan for the following road systems:

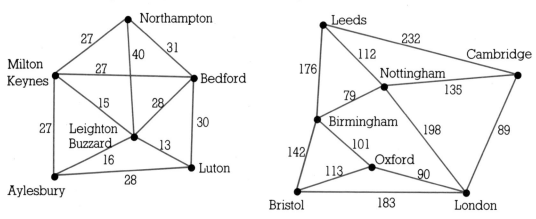

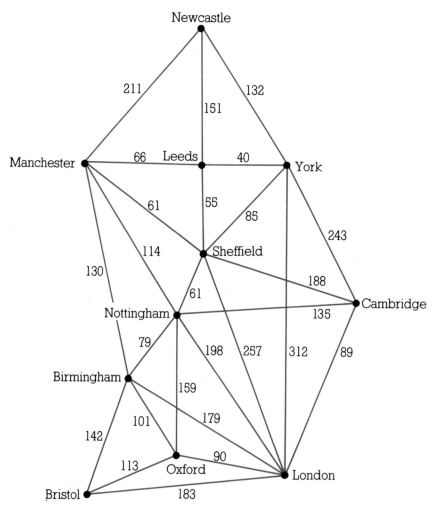

So the key to knowing whether we have to repeat roads or not is the number of roads at each town or vertex. Any vertex with an *even* number of roads will not need any roads being repeated, but any vertex with an *odd* number of roads will require at least 1 road being repeated.

## Question 3

The map below indicates principal railway lines in the Midlands and East Anglia. Find an inspection route which covers every line at least once, but which has the smallest total distance.

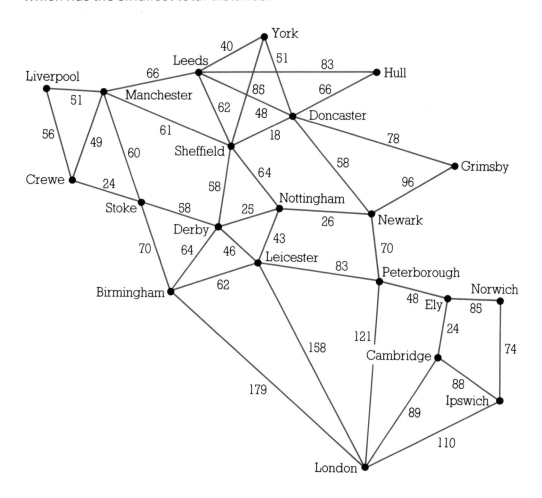

# 23 Frogs 2

For the rules of Frogs, see chapter 13.

## Question 1

Suppose we have $n$ counters on each side in the game of Frogs. Here are 4 possible generalisations for the number of moves needed to complete the game:

$n(n + 2)$
$n^2 + 2n$
$n \times n + n + n$
$(n + 1)^2 - 1$

Show that these are, in fact, all equivalent.

## Question 2

If there are $n$ counters on one side and $m$ on the other the generalisation for the number of moves is:

$$nm + n + m$$

Prove that this becomes any one of the generalisations in question 1 when we make $n = m$.

## Question 3

In this question you are given the number of moves taken to complete 6 different games. Work out the numbers of counters needed on each side ($m$ and $n$) which produced these results. There may be more than 1 answer, or there may be no answer.

| | |
|---|---|
| **1** 29 moves | **4** 80 moves |
| **2** 99 moves | **5** 31 moves |
| **3** 35 moves | **6** 36 moves |

You might consider using a computer program for this question.

## Question 4

Prove that the equation

$$x^2 + 2x = 30$$

has no integer solutions.

## Question 5

Obtain all solutions to the following equations.

| | |
|---|---|
| **1** $n(n + 2) = 80$ | **4** $n^2 + 2n = 63$ |
| **2** $n^2 + 2n = 120$ | **5** $n(n + 2) = 35$ |
| **3** $n^2 + 2n = 440$ | **6** $(n + 1)^2 - 1 = 143$ |

## Question 6

Prove that the equation

$$nm + n + m = 36$$

has no integer solutions where neither $n$ nor $m$ can be zero.

No credit will be given for solutions which are merely a form of substituting in values for $m$ and $n$.

# 24 Cuisenaire Activity 4

## Squares

```
w : white
r : red
g : light green
p : pink
y : yellow
d : dark green
b : black
t : tan (brown)
B : blue
O : orange
```

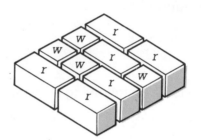

## Question 1

Can you make a square using:

**1** 3 *light green* and 11 *yellow* rods?     **4** 3 *light green* and 10 *pink* rods?

**2** 4 *pink* and 14 *dark green* rods?     **5** 4 *pink* and 13 *yellow* rods?

**3** 6 *dark green* and 16 *pink* rods?

Can you think of a rule which says how many of each colour will be needed in order to make a square?

## Question 2

(i) Find as many ways as you can of joining 3 *light green* and 4 *pink* rods to make a square.

(ii) Make a square using 3 *light green*, 2 *pink*, 4 *yellow* and 2 *dark green* rods.

(iii) Make a square using 3 *light green*, 5 *yellow*, and 5 *dark green* rods.

# Question 3

What size squares can you make using only

(i)  *red* rods
(ii)  *light green* rods?

## Rectangles

## Question 4

Can you make a rectangle with all 10 rods using only 1 rod of each colour (but you cannot simply put all the rods in a straight line)?

Here is the rectangle made with the smallest 3 rods:

It is 2 units wide by 3 units long. Size: 2 × 3.

## Question 5

Try to make rectangles with the following sets of rods. Record your results in a table like the one shown below.

| number of rods | size of rectangle |
|---|---|
| 1 (*w*) | 1 × 1 |
| 2 (*w,r*) | 1 × 3 |
| 3 (*w,r,g*) | 2 × 3 |

**1**  *w r g p*

**2**  *w r g p y*

**3**  *w r g p y d*

**4**  *w r g p y d b*

**5**  *w r g p y d b t*

**6**  *w r g p y d b t B*

**7**  *w r g p y d b t B O*

Can you find patterns in your results? Using your results, could you predict what size the rectangles would be for:

**8**  20 rods

**9**  30 rods

**10**  50 rods

**11**  27 rods?

## Question 6

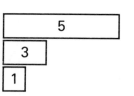

(i) Using 2 sets of these rods make a rectangle. Write down the size of it.

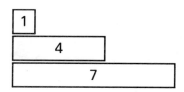

(ii) Do the same with this sequence. Get two sets and make a rectangle. What size is it?

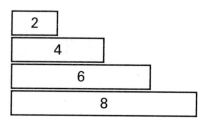

(iii) Using two sets of this sequence, what size rectangle do you get?

Your answer to (i) could have been written:

$$2(1 + 3 + 5) = 3 \times 6$$

Write the others in the same way. Then do question 7.

## Question 7

**1** $2(1 + 4 + 7 + 10) =$

**2** $2(1 + 3 + 5 + 7) =$

**3** $2(2 + 5 + 8) =$

**4** $2(3 + 7 + 11 + 15) =$

**5** $2(5 + 9 + 13 + 17 + 21 + 25) =$

**6** $2(61 + 67 + 73 + 79 + 85) =$

**7** $2(100 + 107 + 114 + 121 + 128 + 135) =$

## Cutting

## Question 6

For this question assume that a cut has no thickness. How can you cut up 5 *blue* rods into one of each of these pieces:

*w, r, g, p, y, d, b, t, B*

# Question 7

Draw diagrams in answering the following questions.

**1** Cut up 4 *orange* rods into these pieces:

$$4 \times r, 9 \times g, y$$

**2** Cut up 3 *orange* rods into these pieces:

$$w, 2 \times r, g, p, y, d, b$$

**3** Cut up 4 *orange* rods into these pieces:

$$w, 4 \times r, g, 3 \times p, b, B$$

**4** Cut up 4 orange rods into these pieces:

$$2 \times r, g, p, d, b, 2 \times t$$

**5** Cut 3 *orange* rods into these pieces:

$$r, 2 \times g, p, y, 2 \times d$$

**6** Cut 5 *orange* rods into these pieces:

$$w, 3 \times r, 2 \times g, 2 \times p, 2 \times d, 2 \times b$$

**7** Cut 5 *orange* rods into these pieces:

$$r, g, p, 4 \times y, t, B$$

# 25 PLC

## Question 1

You will have seen the demonstration of circles touching fixed lines, points and circles. Now do some drawings of your own, on plain paper, in which circles of different sizes touch
(i)   a fixed point (L)
(ii)  a fixed point (P)
(iii) a fixed circle (C).

## Selecting from P, L and C

If we now select any 2 from P, L and C we can draw circles in different positions that touch both.

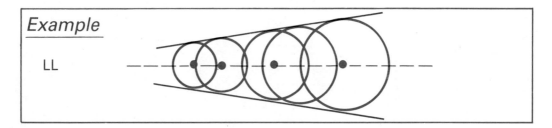

*Example*

LL

## Question 2

You will have seen that the computer traces out the path of the centres of circles moving such that the circles touch any 2 from P, L and C.

With compasses or circle templates, draw your own versions of some or all of the following:

| | |
|---|---|
| **1** PL | **4** LL (intersecting also) |
| **2** PP | **5** CC |
| **3** PC | **6** CL |

In the following questions, only look those situations you have not already considered. Remember that the circles you draw can touch the inside as well as the outside of a circle.

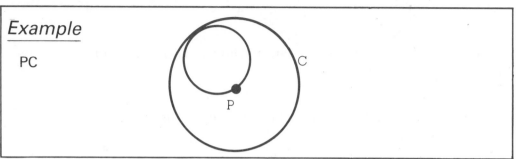

Example

PC

## Question 3

What happens in the following cases for CC?

**1**

circles of different sizes

**4**

overlapping circles of different sizes

**2**

circles close together. What happens when the circles move further apart?

**5**

one circle in another

**3**

overlapping circles of the same size

**6**

concentric circles

## Question 4

Now draw circles that touch both C and L in the following cases:

**1**

line as a tangent

**2**

internal line

**3**

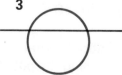

extended line

**4**

line through the centre

**5**

internal line and not touching

## Question 5

Draw circles that touch C and P in the following cases:

**1**

point on the circumference

**2**

internal point

**3**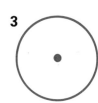

point at the centre

In questions 3 – 5 you should have been able to generate:

straight lines; parabolas; circles; hyperbolae; ellipses and some argument!

Classify which PLC selections give each of the above. Have you found any which give something different?

## Question 6

Can you draw circles for the following selections?

**1** LLL

**2** CCC

**3** PPP

**4** PLC

**5** Other combinations of your own choice.

## Extension problems

## Question 7

Investigate what happens when you extend these ideas to 3 dimensions, this time using the path traced out by the centre of a sphere, relative to planes, points, fixed spheres and lines.

*Example*

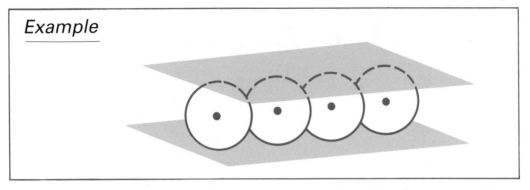

Consider the lines, planes, surfaces or shapes produced by the centres of spheres in the following situations:

**1 point** and a **sphere** (a) point outside sphere; (b) point touching sphere; (c) point inside and at centre of sphere; (d) point inside and not at centre of sphere
**2 point** and a **plane**
**3 intersecting planes**
**4 sphere** and a **plane** (a) plane near sphere; (b) plane touching sphere; (c) plane cutting sphere

# Question 8

Investigate the locus of the mid-point of a line which touches points moving on fixed lines and circles – an example is shown in the diagram below.

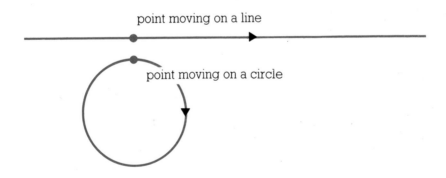

point moving on a line

point moving on a circle

What happens to this locus if the points move
(i) at the same speed
(ii) at different speeds (perhaps one is twice the speed of the other)
(iii) in different directions
(iv) on different fixed objects?

# 26 Pythagoras 3

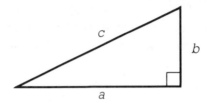

## Question 1

The following sets of whole numbers represent side-lengths of right-angled triangles. They are called Pythagorean triples.

| a | b | c |
|---|---|---|
| 3 | 4 | 5 |
| 5 | 12 | 13 |
| 7 | 24 | 25 |

Can you see a pattern? What would the next 6 triples be in this sequence?

## Question 2

(i) Let

$$a = m^2 - n^2$$
$$b = 2mn$$
$$c = m^2 + n^2$$

Choose values for $m$ and $n$ and work out $a$, $b$ and $c$. Do this for 10 or more different values of $m$ and $n$. You could write a short computer program for this. Check that $a^2 + b^2 = c^2$ for each one.

(ii) If

$$a^2 + b^2 = c^2,$$

    (a) show that $2c^2$ is also the sum of 2 squares.

    (b) For what values of $a$ and $b$ is $c^2$ prime? List some values of $a$ and $b$ that *don't* give prime numbers.

(iii) Copy and complete the table for values of $a$ which are odd numbers.

| $a$ | $b$ | $c$ | $a + b + c$ | factors |
|-----|-----|-----|-------------|---------|
| 3 | 4 | 5 | 12 | $3 \times 4$ |
| 5 | 12 | 13 | 30 | $5 \times 6$ |
| 7 | 24 | 25 | ⋮ | ⋮ |
| 9 | 40 | 41 | ⋮ | ⋮ |
| 11 | | | | |
| 13 | | | | |
| 15 | | | | |
| 17 | | | | |
| 19 | | | | |
| 21 | | | | |

Can you write

    (a) $b$ and $c$ in terms of $a$?
    (b) $a + b + c$ in terms of $a$?
    (c) factors in terms of $a$?

The final 4 questions are more open-ended than the other questions in this chapter. They are designed for you to investigate in your own way.

## Question 4

Pythagoras' theorem states that for any right-angled triangle with squares $A$, $B$ and $C$ drawn on its sides:

    area $A$ + area $B$ = area $C$

  Is the theorem true only for squares or could we draw other shapes on the sides?

  Illustrate the report of your investigation with appropriate diagrams, calculations and explanations.

## Question 5

Pythagoras' theorem is essentially for triangles. Is there any similar result for quadrilaterals (or larger polygons)?

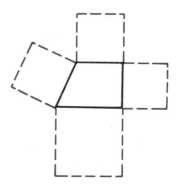

Be careful to write an accurate account of everything you try even if it is not successful. Show all your working and clearly state any conclusions you make.

## Question 6

The most famous Pythagorean triple is 3, 4, 5:

$$3^2 + 4^2 = 5^2$$

Because $3^2 + 4^2 = 5^2$ it follows that $6^2 + 8^2 = 10^2$, $9^2 + 12^2 = 15^2$, $12^2 + 16^2 = 20^2$ etc. Show that the sets of numbers (3, 4, 5), (6, 8, 10), (9, 12,15), (12, 16, 20), etc. are the *only* sets of Pythagorean triples where the 3 numbers are in **arithmetic progression**.

## Question 7

Use the symbolic forms in question 2 to help with the solution to the **peculiar triangle problem**, in which some right-angled triangles have area and perimeter that are numerically equal.

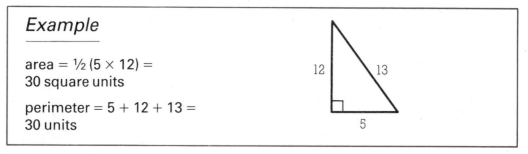

### Example

area = ½ (5 × 12) =
30 square units

perimeter = 5 + 12 + 13 =
30 units

You have to find more of these right-angled triangles. Can you generalise which triangles have numerically equal area and perimeter?

# 27 Transformations

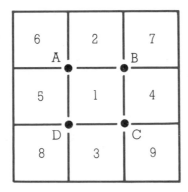

Rotations of 180° about points A, B, C or D in the diagram, are written as

$$R_A, R_B, R_C, R_D$$

Reflections in lines AB, BC, CD, DA, AC, BD are written as:

$$M_{AB}, M_{BC}, M_{CD}, M_{DA}, M_{AC}, M_{BD}$$

Always carry out the rotation or reflection nearest the bracket first e.g.
$R_B M_{AC} (1)$, do $M_{AC}$ first then $R_B$.

## Question 1

Using the diagram above find the number which is the result of these
rotations and reflections?

**1** $R_A (1) =$

**2** $R_B (1) =$

**3** $R_C (4) =$

**4** $R_D (8) =$

**5** $M_{AB} (1) =$

**6** $M_{AB} (4) =$

**7** $M_{CD} (8) =$

**8** $M_{DA} (8) =$

**9** $M_{AC} (8) =$

**10** $M_{BD} (8) =$

**11** $R_A M_{AB} (1) =$

**12** $M_{DA} M_{CD} (1) =$

**13** $R_B R_C (9) =$

**14** $R_A R_D R_C (4) =$

**15** $M_{AC} M_{AC} (1) =$

**16** $M_{AC} M_{DA} (1) =$

**17** $M_{BD} R_C (4) =$

**18** $M_{AB} M_{CD} M_{BD} (6) =$

**19** $M_{AC} R_D (3) =$

**20** $M_{BD} R_D R_D R_C (4) =$

**21** $M_{AB} R_B (?) = 1$

**22** $M_{CD} R_A (?) = 8$

**23** $M_{AB} M_{BD} (?) = 5$

**24** $R_B M_{AC} (?) = 2$

## Question 2

In the next question, all rotations are around the centre point.

$R_1$ = clockwise rotation of 120°

$R_2$ = clockwise rotation of 240°

You may find tracing paper helpful for this question.

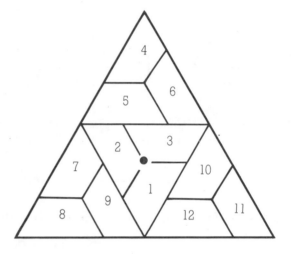

---

### Example

$$R_1 (1) = 2$$

---

Now try these:

**1** $R_1 (2) =$

**2** $R_1 (3) =$

**3** $R_1 (3) =$

**4** $R_2 (1) =$

**5** $R_1 (5) =$

**6** $R_1 (6) =$

**7** $R_1 (7) =$

**8** $R_1 (8) =$

**9** $R_2 (10) =$

**10** $R_2 (9) =$

**11** $R_2 (11) =$

**12** $R_2 (7) =$

**13** $R_1 (12) =$

**14** $R_1 (?) = 5$

**15** $R_2 (?) = 10$

**16** $R_1 (?) = 7$

**17** $R_2 (?) = 4$

**18** $R_1 (?) = 6$

**19** $R_2 (?) = 2$

**20** $R_1 (?) = 11$

## Question 3

You could use tracing paper to make a copy of the diagram in question 2 to show where you would put a centre of rotation to turn:

**1** $3 \to 6$            **3** $10 \to 2$

**2** $3 \to 5$            **4** $8 \to 2$

## Question 4

$R_A$ means 180° rotation about the point A.
$R_B$ means 180° rotation about B, etc.

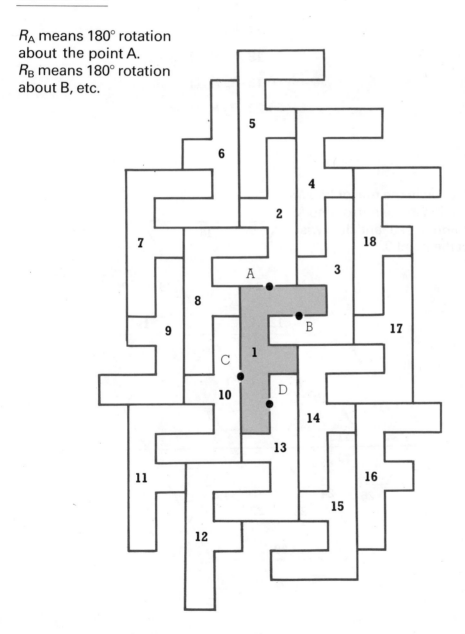

Find the results of these rotations. Remember to do the rotation nearest the bracket first. Again you may find tracing paper helpful.

**1** $R_A (1) =$

**2** $R_B (1) =$

**3** $R_C (1) =$

**4** $R_D (1) =$

**5** $R_A (3) =$

**6** $R_B (8) =$

**7** $R_C (14) =$

**8** $R_D (10) =$

**9** $R_A (5) =$

**10** $R_B (2) =$

**11** $R_C (12) =$

**12** $R_D (3) =$

**13** $R_A (9) =$

**14** $R_B (10) =$

**15** $R_C (9) =$

**16** $R_A R_B (1) =$

**17** $R_B R_A (1) =$

**18** $R_C R_A (1) =$

**19** $R_B R_D (1) =$

**20** $R_D R_C (1) =$

## Question 5

$M_1$: a reflection in the mirror line $M_1$
$M_2$: a reflection in the mirror line $M_2$
$R$: a rotation of 120° anticlockwise about the point 0.

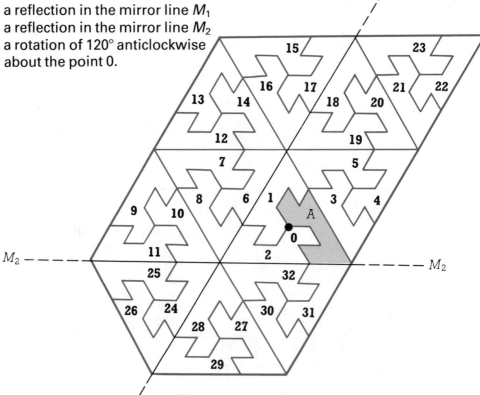

$$M_2 (A) = 31$$
$$M_1 (31) = 9$$

so $\quad M_1 M_2 (A) = 9$

Now try these remembering to do the transformation nearest the bracket first.

**1** $M_1 (A) =$

**2** $M_1 (5) =$

**3** $M_1 (27) =$

**4** $M_1 (11) =$

**5** $M_1 (14) =$

**6** $M_2 (2) =$

**7** $M_2 (25) =$

**8** $M_2 (28) =$

**9** $M_2 (29) =$

**10** $M_2 (1) =$

**11** $R (A) =$

**12** $R (2) =$

**13** $R (3) =$

**14** $R (6) =$

**15** $R (29) =$

**16** $M_1 M_1 (A) =$

**17** $M_2 M_1 (A) =$

**18** $M_1 R (A) =$

**19** $M_2 R (A) =$

**20** $R R (A) =$

**21** $M_1 M_2 M_1 (A) =$

**22** $M_2 M_1 R (A) =$

**23** $R M_1 (A) =$

**24** $M_2 R M_1 (A) =$

How many positions can you reach from (A) using the transformations $M_1$, $M_2$ and $R$ in any combination?

## Question 6

If two identical shapes were dropped onto a flat surface, would it always be possible to move from one position to another using only reflections? Can you find any positions that are impossible to reach by a combination of reflections?

# 28 Groups 2

## Symmetry Groups

You may find it helpful to use a labelled cut-out rectangle for this question.
The 4 elements in the symmetry group of a rectangle are:

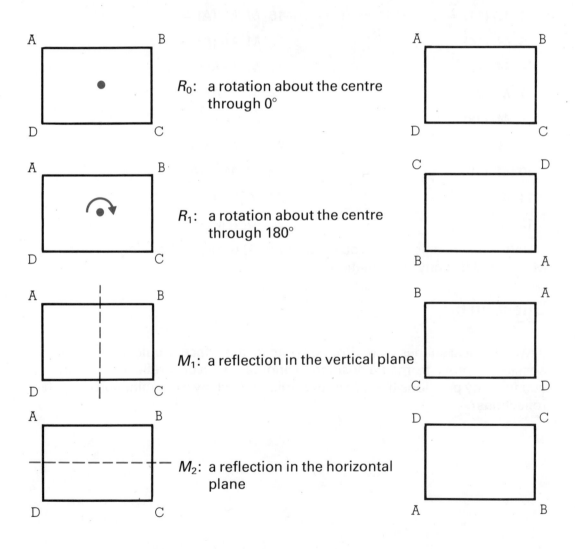

$R_0$: a rotation about the centre through 0°

$R_1$: a rotation about the centre through 180°

$M_1$: a reflection in the vertical plane

$M_2$: a reflection in the horizontal plane

# Question 1

How many elements are there in the symmetry groups of the following shapes. Illustrate your answer with appropriate diagrams.

**1**  a square

**2**  an isoceles triangle

**3**  an equilateral triangle

**4**  a regular pentagon

**5**  a regular hexagon

**6**  the letter H

**7**  a circle

**8**  a perfect ellipse?

# Question 2

A square is a regular polygon with 4 sides, whilst an equilateral triangle is a regular polygon with 3 sides.

What is the relationship between the number of sides of a regular polygon and the number of elements in its symmetry group? Express your results in symbols and illustrate your answer with appropriate diagrams.

# Question 3

How many elements are there in the symmetry groups of each of these 3-dimensional shapes:

**1**  a sphere

**2**  a regular tetrahedron

**3**  a regular octahedron

**4**  a cube

**5**  2 other 3-dimensional shapes of your own choosing?

You should illustrate your answers through the use of appropriate 3-dimensional diagrams.

## Operation tables

We can combine the elements of the rectangle symmetry group by doing, for instance, $R_1$ followed by $M_2$:

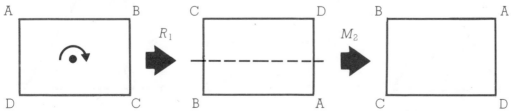

115

$R_1$ followed by $M_2$ gives the same as $M_1$ i.e. $R_1 \bullet M_2 = M_1$

## Question 4

Complete the **operation table**:

| | 2nd operation | | | |
|---|---|---|---|---|
| $\bullet$ | $R_0$ | $R_1$ | $M_1$ | $M_2$ |
| $R_0$ | | | | |
| $R_1$ | | | $M_1$ | |
| $M_1$ | | | | |
| $M_2$ | | | | |

1st operation

## Question 5

Solve the equations

**1** $R_1 \bullet x = M_2$

**2** $x \bullet M_2 = M_1$

**3** $R_1 \bullet x = R_0$

**4** $x \bullet M_1 = R_1$

## Question 6

You will find it helpful to cut out an equilateral triangle and label the vertices on the front and the back for this question.
The 6 elements in the symmetry group of an equilateral triangle are:

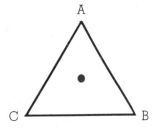

$R_0$: rotation through $0°$

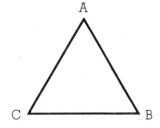

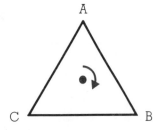

$R_1$: rotation through $120°$

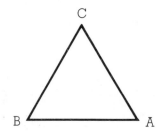

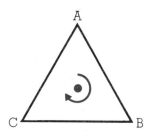

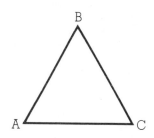

$R_2$: rotation through 240°

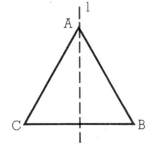

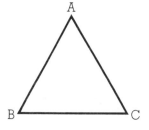

$M_2$: reflection in line 1

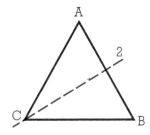

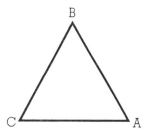

$M_2$: reflection in line 2

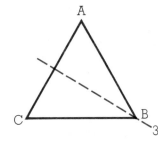

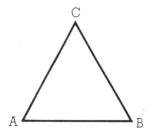

$M_3$: reflection in line 3

Complete the following operation table:

| • | $R_0$ | $R_1$ | $M_1$ | $M_2$ | $M_3$ |
|---|---|---|---|---|---|
| $R_0$ | | | | | |
| $R_1$ | | | | | |
| $M_1$ | | | | | |
| $M_2$ | | | | | |
| $M_3$ | | | | | |

117

## Identity and inverse elements

There are some specific terms which relate to groups and operation tables in general. These are:

(i) **Identity element**
   This is such that

$$\begin{bmatrix} \text{an element you} \\ \text{start with} \end{bmatrix} \bullet \begin{bmatrix} \text{identity} \\ \text{element} \end{bmatrix} = \begin{bmatrix} \text{the element you} \\ \text{start with} \end{bmatrix}$$

   i.e. it does not change anything.

(ii) **Inverse element**
   This is such that

$$\begin{bmatrix} \text{an element you} \\ \text{start with} \end{bmatrix} \bullet \begin{bmatrix} \text{its inverse} \\ \text{element} \end{bmatrix} = \begin{bmatrix} \text{the identity} \\ \text{element} \end{bmatrix}$$

## Question 7

Using your results, for the symmetry group of the equilateral triangle find

(i) the identity element

(ii) the inverse elements of each of the 6 elements.

> An operation is called **commutative** if in *all cases*
>
> [first element] • [second element] = [second element] • [first element]

## Question 8

(i) Is the operation of combining the elements of the symmetry group of the equilateral triangle commutative? You will need to check all the possible combinations — to be commutative it needs to work in *all* cases.

(ii) If the operation is commutative, show 3 separate combinations which demonstrate this. If it is not commutative give 1 example which shows this case.

We shall write

$$R_1 \bullet R_1 \text{ as } R_1{}^2$$

and

$$M_1 \bullet M_1 \bullet M_1 \bullet M_1 \bullet M_1 \text{ as } M_1{}^5$$

# Question 9

Simplify as much as possible

  **1** $R_2{}^5$                          **2** $M_2{}^3$

  **3** $R_1{}^8$                          **4** $R_2{}^7 \bullet M_2{}^5 \bullet M_1{}^4$

Invent and then simplify some more versions of your own.

# Question 10

This question is about the symmetry group of a square.

  (i) Define the 8 elements in the symmetry group of a square.

 (ii) Set up a group operation table.

(iii) Define the identity element of this group.

(iv) Also define the inverse of each of the 8 elements.

 (v) Is this group commutative? Illustrate your answer with examples.

# 29 The Sigma Function

This chapter looks at some of the properties of the Sigma function $\sigma(n)$, which adds together the divisors of a number, and of the function we call $\tau(n)$ which gives the total number of such divisors.

## Question 1

First copy and complete the following table as accurately as you can. In order to have some information on which to work, some of the entries have been completed for you. Fill in the table up to $n = 30$, at least.

| $n$ | divisors of $n$ | number of divisors ($\tau(n)$) | sums of divisors ($\sigma(n)$) |
|---|---|---|---|
| 1 | 1 | 1 | 1 |
| 2 | 1, 2 | 2 | 3 |
| 3 | 1, 3 | 2 | 4 |
| 4 | 1, 2, 4 | 3 | 7 |
| 5 | | | |
| 6 | | | |
| 7 | | | |
| 8 | | | |
| 9 | | | |
| 10 | 1, 2, 5 | 3 | 8 |
| etc. | | | |
| 30 | | | |

Consider $n = 12$: since 12 has the divisors 1, 2, 3, 4, 6, 12, we find that

$$\tau(12) = 6 \text{ and}$$
$$\sigma(12) = 1 + 2 + 3 + 4 + 6 + 12 = 28$$

From the table we see that, for example,

$$\tau(1) = 1, \tau(2) = 2, \tau(3) = 2, \tau(4) = 3 \text{ etc.}$$
$$\text{and } \sigma(1) = 1, \sigma(2) = 3, \sigma(3) = 4, \sigma(4) = 7 \text{ etc.}$$

# Question 2

What can you say about $\sigma(n)$ and $\tau(n)$ when $n$ is a prime number?

# Question 3

A **perfect number** is one for which the sum of the divisors (but *not* including the number itself) are equal to the number itself.
   Which are the perfect numbers in your list?

# Question 4

A **Mersenne prime number** is one which can be expressed in the form $2^n - 1$ for some value of $n$. Check that 3 and 7 are Mersenne primes. Now check that the formula

$$\frac{p(p + 1)}{2}$$

gives perfect numbers when $p$ is a Mersenne prime. Using this result, can you find the next perfect number?

## Prime-factor form

To express a number in prime-factor form, divide first by 2 until you get an odd number, then try to divide by 3, then 5 etc. The divisors are always prime-numbers.

---

*Example*

$$\begin{array}{r} 2)\overline{36} \\ 2)\overline{18} \\ 3)\overline{9} \\ 3)\overline{3} \\ 1 \end{array}$$

So the prime-factors are $2 \times 2 \times 3 \times 3$ or

$$2^2 \times 3^2.$$

---

## Question 5

Complete the following table:

| number $n$ | prime-factor form |
|:---:|:---:|
| 2 | $2^1$ |
| 3 | $3^1$ |
| 4 | $2^2$ |
| 5 | $5^1$ |
| 6 | $2^1 \times 3^1$ |
| etc. to $n = 30$, at least. | |

Consider $n = 36$: it has the following 9 divisors:

$$1, 2, 3, 4, 6, 9, 12, 18, 36.$$

So we write

$$\tau(36) = 9$$

But from the example before question 5 we see that

$$36 = 2^2 \times 3^2$$

and that by *adding 1 to each of the powers and then multiplying them*, we obtain $3 \times 3 = 9$ also.

## Question 6

Does this work in the following cases:

**1** $n = 24$

**2** $n = 20$

**3** $n = 7$

**4** $n = 48$

**5** $n = 100$

**6** Try some of your own as well.

## Question 7

Can you show whether it will always work? In other words, can you prove it?

From your tables you will see that:

$$\tau(20) = 6 \quad \text{and} \quad \sigma(20) = 42$$

Trying to simplify the problem by setting $20 = 2 \times 10$ we have:

$$\tau(2 \times 10) = \tau(2) \times \tau(10)$$
$$= 2 \times 4$$
$$= 8 \text{ (wrong number)}$$

and

$$\sigma(2 \times 10) = \sigma(2) \times \sigma(10)$$
$$= 3 \times 18 = 56 \text{ (also wrong)}$$

But if we set $20 = 4 \times 5$ we have:

$$\tau(4 \times 5) = \tau(4) \times \tau(5)$$
$$= 3 \times 2 = 6 \text{ (correct answer)}$$

and

$$\sigma(4 \times 5) = \sigma(4) \times \sigma(5)$$
$$= 7 \times 6 = 42 \text{ (also correct)}$$

## Question 8

Try to identify what happens in the following cases:

**1** $n = 24$   using $(12 \times 2)$
then $(6 \times 4)$
then $(8 \times 3)$

**2** $n = 30$   using $(10 \times 3)$
then $(5 \times 6)$

**3** $n = 28$   using $(14 \times 2)$
then $(7 \times 4)$

**4** Now think of some values for $n$ and do the same as you have for 1 – 3. Try to work out the circumstances under which

$$\tau(n \times m) = \tau(n) \times \tau(m)$$
$$\text{and } \sigma(n \times m) = \sigma(n) \times \sigma(m)$$

## Question 9

The following problem is based on one from a high level Mathematics question paper.

(i) If $p =$ any prime number, can you prove

(a) $\tau(p^n) = n + 1$

(b) $\sigma(p^n) = \dfrac{p^{n+1} - 1}{p - 1}$

(ii) Try this result in the following cases:

$$2^2, 2^3, 2^4, 4^2, 2^5, 3^2, 6^2, 3^3, \ldots \text{etc.}$$

(iii) Using the result from question 8, evaluate

$$\sigma(15), \sigma(540), \sigma(600).$$

## Oddities

## Question 10

Is it true that the *product* of the divisors of a number $(n) = n^{\tau(n)/2}$?

## Question 11

Is $\tau(n)$ an odd number only when $n$ is a perfect square *or* twice a perfect square?

# 30 Manhattan Routes

In the town of Manhattan, the streets are all straight and viewed from above create a square grid like this:

Suppose 2 points, A and B are marked at the corners on the grid:

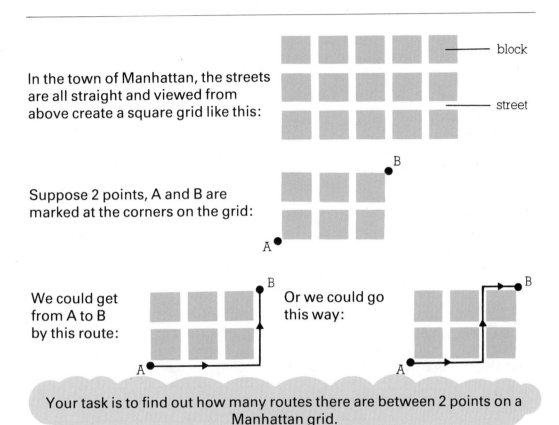

We could get from A to B by this route:

Or we could go this way:

Your task is to find out how many routes there are between 2 points on a Manhattan grid.

*Rule 1* You start at a point which is never above nor to the right of the finish point.

*Rule 2* You can only move along the streets or grid lines.

*Rule 3* Your start and finish points must be at corners where grid lines cross.

*Rule 4* You can only move to the right and upwards — never to the left or down.

## Question 1

With these rules, and for the 2 points A and B above there are 10 different routes. Draw diagrams to show all 10 routes.

# Question 2

In each of the cases below, find the total number of routes from the start to the finish and investigate further.

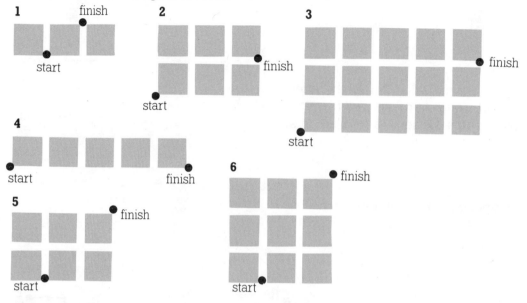

## Manhattan Loci

At the beginning of the chapter, one solution of the locus A B = 5 was shown. For the following questions, change *rule 4* to 'You must finish as far from the starting point as possible.' Here is an example to illustrate the new rule:

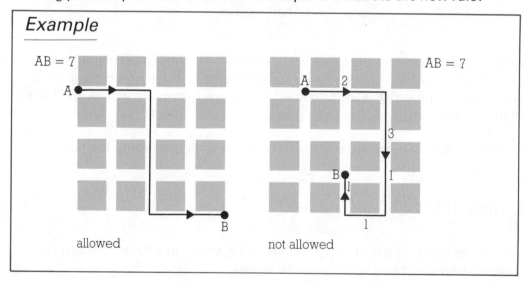

# Question 3

Select a fixed point A on your grid and then plot on separate diagrams the locus of

**1** A B = 7            **2** A B = 2            **3** A B = 4            **4** A B = 7

# Question 4

What can you say about the locus A B = $n$ units?

# Question 5

Investigate what happens if we ignore *rule 4* for question 4.

# Question 6

The diagram shows A and B, 6 blocks apart. $P_1$ and $P_2$ satisfy the locus AP = PB.

(i) Complete the locus and describe it in words.
(ii) Investigate how the locus changes if
   (a) A and B are 3 blocks apart
   (b) A and B are 4 blocks apart on the *same* street.

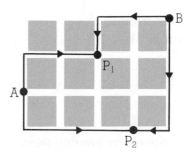

# Question 7

A and B are fixed points. Investigate the locus of P produced by the following relationships

**1** AP = 2PB       **2** AP = 3PB       **3** AP = 4PB       **4** AP = ½ PB

# Question 8

In all of these, A and B are on the same street. Find and describe where you can, the locus if

**1** AP = PB           **3** AP + PB = 10

**2** AP = 8            **4** AP × PB = 12

# 31 Turning Points

## Quadratic graphs

The general form of the quadratic graph is:
$$y = ax^2 + bx + c$$

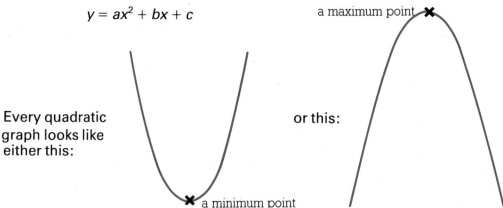

a maximum point

Every quadratic graph looks like either this:

or this:

a minimum point

These both show **turning points**.

Your task is to find the co-ordinates of the turning points. You need to be able to find the relationship between these co-ordinates and the equation of the graph.

To help you with your investigation it is suggested that you do the following things with your quadratics.

(i) Sketch the graph.

(ii) Substitute in numbers, possibly using a calculator or computer.

(iii) Complete the square.

### Example

$$y = x^2 - 4x + 9$$

**1** Graph

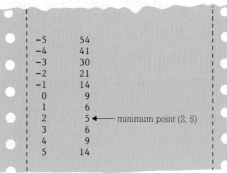

**2** Computer program

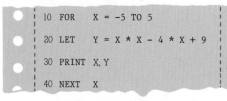

```
10  FOR    X = -5 TO 5
20  LET    Y = X * X - 4 * X + 9
30  PRINT  X, Y
40  NEXT   X
```

```
-5        54
-4        41
-3        30
-2        21
-1        14
 0         9
 1         6
 2         5  ←──── minimum point (2, 5)
 3         6
 4         9
 5        14
```

**3** By completing the square

$$x^2 - 4x + 9$$

We can complete the square by the method shown in chapter 17, Quadratic Equations. This gives:

$$x^2 - 4x + 9$$
$$= (x - 2)^2 - 4 + 9$$
$$= (x - 2)^2 + 5$$

From the examples shown already, the minimum point is (2, 5). Look again at the result of completing the square.

$$x^2 - 4x + 9 = (x - 2)^2 + 5$$
$$\text{minimum point (2, 5)}$$

Do you notice something special?

## Question 1

Here are 10 different quadratics for you to work on. Find the turning points using the methods shown in the example. To be sure of your answer, use two methods on a few of the equations.

**1** $y = x^2 - 4x - 3$      **4** $y = x^2 + 6x + 11$

**2** $y = x^2 + 4x - 7$      **5** $y = x^2 - 8x + 15$

**3** $y = 5 + 2x - x^2$      **6** $y = 9 + 8x - x^2$

**7** $y = 2x^2 - 12x + 3$

**8** $y = 7 + 8x - 2x^2$

**9** $y = 3x^2 - 18x + 15$

**10** $y = 4x^2 + 24x + 11$

## Question 2

Obtain the co-ordinates of the turning point of the graphs:

**1** $y = x^2 - 6x + 7$

**2** $y = 11 + 4x - x^2$

**3** $y = 2x^2 - 10x + 17$

**4** $y = 2x^2 + 9x - 7$

State whether each one is a maximum or a minimum.

# 32 Iteration

It will be possible to answer the following questions using 'trial and error' methods. You will need a calculator for all these questions.

## Question 1

Find the numbers $x$ and $y$ such that

$$x + y = 79$$
$$\text{and} \quad x - y = 23$$

## Question 2

Find $x$ and $y$ in each of the following cases:

**1** $x + y = 100$ and $xy = 2499$

**2** $x + y = 100$ and $xy = 1971$

**3** $x + y = 100$ and $xy = 2475$

**4** $x + y = 100$ and $xy = 2244$

**5** $x + y = 100$ and $xy = 1411$

**6** $x + y = 100$ and $xy = 2379$

## Question 3

**1** The product of 2 consecutive numbers is 545382. Find them.

**2** The product of 3 consecutive numbers is 42840. Find them.

## Question 4

**1** Find a number $x$ such that $x^3 = 10$

**2** Find a number $x$ such that $x^2 + x = 10$

**3** Find 2 different numbers $x$ such that $2^x = x^2$

**4** Find a number $x$ such that $x^4 = 4^x$

## Question 5

You are allowed six successive guesses to find the square root of each of the following. Only use the square root function on the calculator at the end to check the accuracy of your result.

**1** 361

**2** 841

**3** 529

**4** 729

**5** 4096

**6** 3844

## Question 6

The result of dividing 2 single-digit numbers is 0.8571428. What are they?

## Question 7

Find the numbers in the circles so that they add up to the numbers in the squares.

**1**
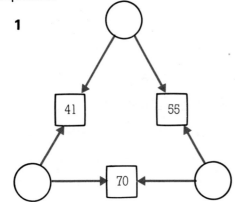

**2**
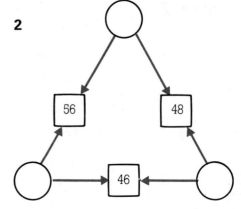

## Question 8

Find a number $x$ such that

$$5.0 < \frac{27}{x} < 5.1$$

## Question 9

Find $a, b, c$ (each of which is a 2-digit number) such that

$$a \times b = 266, b \times c = 209, \text{ and } a \times c = 154.$$

The method of guessing, then making a better guess and systematically working towards the answer, is called an iterative process or an **iteration**.

## Hero's method

Here is an example of an old method for finding square roots by iteration, it is called Hero's method.

---

### Example

To find $\sqrt{60}$
   First guess: $x_1 = 6$

$$\frac{60}{x_1} = 10 \qquad \text{Next guess is half way between 6 an 10.}$$

Second guess: $x_2 = 8$

$$\frac{60}{8} = 7.5 \qquad \text{Next guess is halfway between 8 and 7.5.}$$

   Third guess: $x_3 = 7.75$

$$\frac{60}{7.75} = 7.7419354 \qquad \text{Next guess is half way between these and so on . . .}$$

---

## Question 10

(i) Use this method to work out the square roots of

**1** 17                  **3** 59

**2** 28                  **4** 176

(ii) Use this method to work out the square root of 117 starting with $x_1 = 1$

## Question 11

Write a computer program that will find square roots by Hero's methods to any required degree of accuracy.

# 33 Frogs 3

We can play games very similar to Frogs but with slightly different rules. Here are a few examples.

## Spokes

### Question 1

With 3 spokes, 2 counters in each space, you have to move the counters about to give the result shown in the diagram.

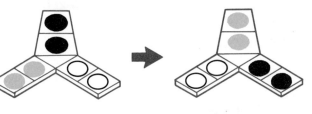

It is up to you to make up the rules, but they should be similar to those in the basic Frog game. You will have to have some rule about jumping around the central triangle. Think carefully about this rule.

Try to find a **generalised result** for the number of moves needed to complete the game with *n* spokes and *m* counters in each spoke.

## Sets of 2 counters

### Question 2

In this game you have to move 3 sets of 2 counters around the board, again using rules of your own.

Start like this:

and finish like this:

## Question 3

Start with the arrangement shown on the left and move the counters to produce the arrangement shown on the right.

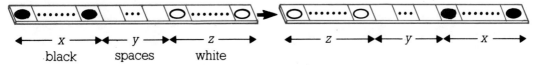

Choose your rules carefully and decide on the size of the spaces $x$, $y$ and $z$.

## Boards of different shapes

## Question 4

Move the counters to get from the arrangement on the left to the one on the right.

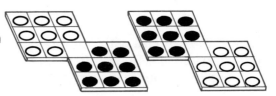

# 34 Billiards 1

## Paths and exit corners

### Question 1

The picture shows a billiard table, 1 ball and the direction of the cue. Copy the picture and show the path taken by the ball after being struck. (No spin, trick shots, etc).

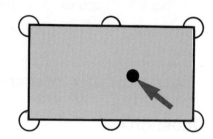

### Question 2

Repeat the exercise with 3 different starting points and aim the cue in any direction you like. Remember that all billiard tables are designed so that the longer side is double the length of the shorter.

## Different tables

We are going to change things for this investigation. You will make up some table sizes of your own and look at the path taken by a single ball. Also we will only have corner pockets in the tables.

In every case there is just 1 ball on the table, it starts in the bottom right-hand corner, marked D, and moves at 45° to the sides of the table. Please accept that it is always hit hard enough to complete its path. We have marked the table like a grid.

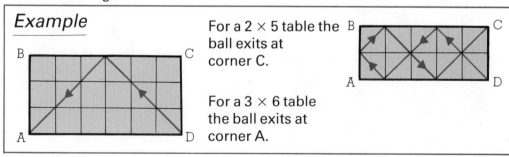

*Example*

For a 2 × 5 table the ball exits at corner C.

For a 3 × 6 table the ball exits at corner A.

## Question 3

| See what happens for tables of | Then try to predict what happens for | And then |
|---|---|---|
| $1 \times 6$ | $7 \times 6$ | $1 \times 12$ |
| $2 \times 6$ | $8 \times 6$ | $2 \times 12$ |
| $3 \times 6$ | $12 \times 6$ | $3 \times 12$ |
| $\vdots$ | $17 \times 6$ | $\vdots$ |
| up to a $6 \times 6$ table. | and $60 \times 6$ tables. | $12 \times 12$ tables and $n \times 12$ tables. |

## Question 4

(i) Will the ball always finish in one of the corners? Remember, you must assume it is always hit hard enough to be able to finish.

(ii) Can a table be designed so that a ball finishes in the corner it starts in?

(iii) Can a table be designed so that a ball can bounce around indefinitely?

(iv) Given that you know the size of the table, can you predict the corner through which it will exit? This might be written in a **generalised** form. Try to do it.

(v) Could you *predict* the exit corners for tables of size:

**1** $3 \times 10$      **4** $2 \times 100$

**2** $6 \times 15$      **5** $56 \times 1001$

**3** $14 \times 27$

# 35 Travelling Salesman

## Introduction

In this investigation, we will try to find the best route for sales people or delivery vans who have to visit a number of places. For example a bread delivery van starts from the bread making factory, delivers bread to a number of shops and returns to the factory. To save petrol and time, the bread firm will want to find the route which has the shortest total distance.

---

### Example

Each day a bread delivery van starting from Sheffield has to deliver bread to Aston, Rotherham and Chapletown returning to Sheffield. The distances in kilometres along the roads between these towns is shown below.

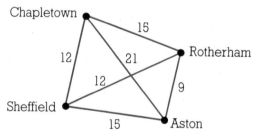

We want to find the route, starting and finishing at Sheffield, which passes through the 3 towns, and has shortest total length. There are 3 possible routes:

S − C − R − A − S   length   12 + 13 + 9 + 15 = 49 km
S − C − A − R − S             12 + 21 + 9 + 12 = 54 km
S − R − C − A − S             12 + 13 + 21 + 15 = 61 km

Therefore the best route is the first one.

---

Now try the following questions. In each case we want to find the route of smallest total length passing through *all* the places marked.

# Question 1

Start and finish at Bedford.

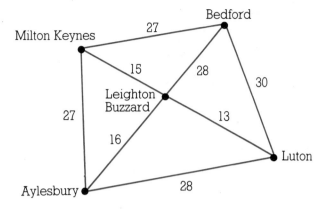

# Question 2

Start and finish at Worthing.

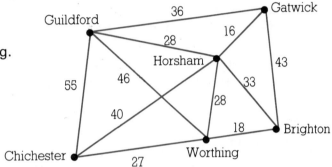

# Question 3

Start and finish at London.

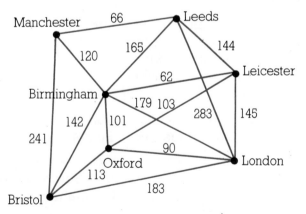

As the network gets more complicated, it gets more difficult to solve the problem. You can easily find routes passing through the places, but are you sure that you have the route of shortest possible length?

There are no exact methods for finding the best solution, but in the next example is a method which usually gets very close to the best solution.

## Example

Start and finish at Birmingham.

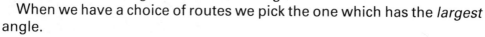

We start by joining all the outside edges (shown in black) so we have the route:

Manchester – Hull – Cambridge – London – Southampton – Bristol – Manchester.

We only have 2 further cities to add into the route. We consider each one, and measure the angle subtended by joining the city to neighbouring cities on the outer route (provided that the roads exist). So for Birmingham there is only 1 possible route.

Manchester – Birmingham – Bristol — angle 150°.

When we have a choice of routes we pick the one which has the *largest* angle.

So we now have the route:

Birmingham – Manchester – Hull – Cambridge
London – Southampton – Bristol – Birmingham.

The last city to be included is Leicester, and we again note the angles subtended.

| | |
|---|---|
| London – Leicester – Cambridge | — angle 32° |
| Cambridge – Leicester – Manchester | — angle 97° |
| Hull – Leicester – Manchester | — angle 60° |
| Manchester – Leicester – Birmingham | — angle 70° |
| Birmingham – Leicester – Bristol | — angle 40° |

The largest angle is 97°, and so we bring in the route Cambridge – Leicester – Hull instead of Cambridge – Hull. So our final route is:

Birmingham – Manchester – Hull – Leicester –
Cambridge – London – Southampton – Bristol –
Birmingham

which has total length

120 + 144 + 133 + 103 + 82 + 115 + 111 + 132 = 940 kilometres

Although we cannot be sure that this is the shortest route, we can be sure that we will not find any route very much shorter. Can you find a route shorter than 940 kilometres?

# Question 4

Start and finish in London and find a route round all the places marked which has a total distance as short as possible.

This investigation finishes with a number of questions which show the many variations of this problem.

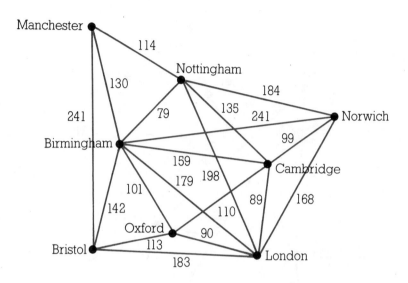

# Question 5

A chain of fish shops is supplied from the docks at Hull. They have two delivery vans. If deliveries are to be made at all the places listed below, plan the two routes so that the total distance travelled is as small as possible. All the distances are in kilometres.

| | | | | | | | |
|---|---|---|---|---|---|---|---|
| Birmingham | 165 | 292 | 198 | 120 | 307 | 75 | 115 |
| Bradford | 163 | 97 | 51 | 144 | 112 | 55 | |
| Carlisle | 229 | 177 | 85 | 271 | 217 | | |
| Hull | 142 | 181 | 135 | 97 | | | |
| Manchester | 198 | 94 | 57 | | | | |
| Newcastle | 235 | 187 | | | | | |
| Nottingham | 55 | | | | | | |
| Sheffield | | | | | | | |

# Question 6

The managing director of the Sigma group of department stores is planning her annual inspection of her stores. The locations of the stores are shown on the map together with the distances between them. Plan a route for her starting and finishing at London, which is as short as possible.

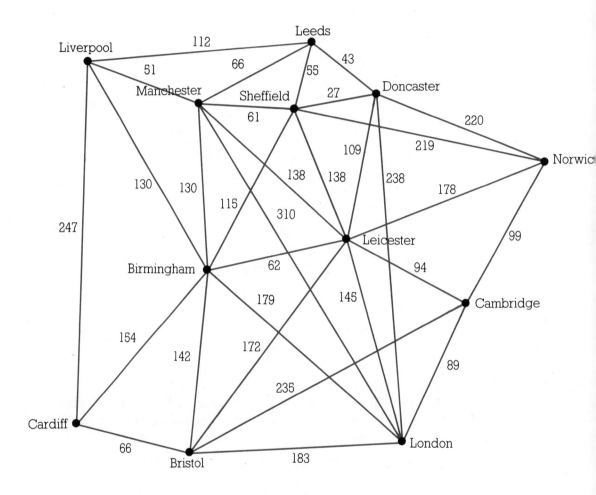

# Question 7

Plymouth City Transport is to start a new minibus service passing through the railway station, the bus and coach station, West Hoe Pier, the market, Royal Parade, the Barbican and the Polytechnic.

Design a route for the minibus with a total route length as short as possible.

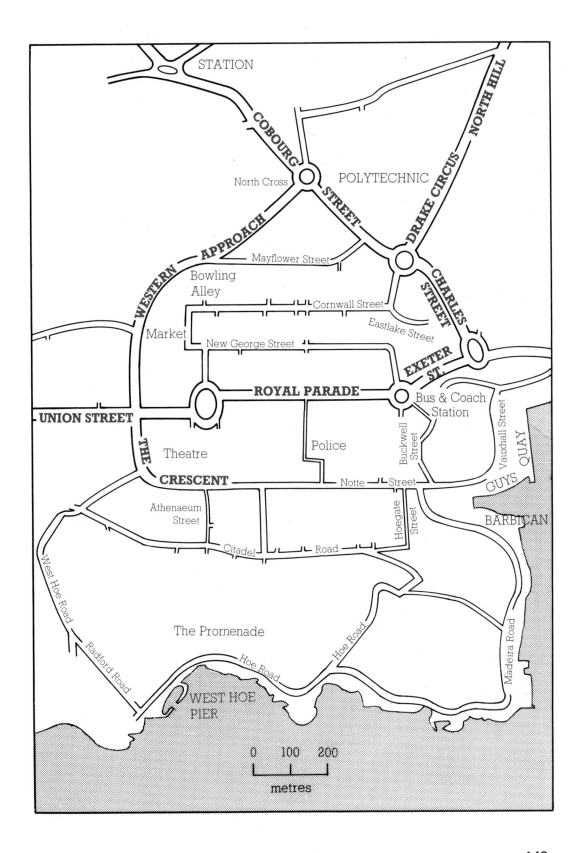

STATION

COBOURG STREET

NORTH HILL

North Cross

POLYTECHNIC

DRAKE CIRCUS

WESTERN APPROACH

Mayflower Street

CHARLES STREET

Bowling Alley

Cornwall Street

Eastlake Street

Market

New George Street

EXETER ST.

ROYAL PARADE

Bus & Coach Station

Vauxhall Street

QUAY

UNION STREET

THE

Theatre

Police

Buckwell Street

GUYS

CRESCENT

Notte Street

BARBICAN

Athenaeum Street

Hoegate Street

Citadel Road

West Hoe Road

Radford Road

The Promenade

Hoe Road

Hoe Road

Madeira Road

Hoe Road

WEST HOE PIER

0   100   200

metres

143

# 36 Co-Prime Investigation

2 numbers are co-prime if the only integer which goes into, or is a factor, or is a divisor, of both of them is 1.

## Question 1

State whether or not these pairs of numbers are co-prime.

**1** 3 and 7

**2** 4 and 6

**3** 5 and 23

**4** 14 and 27

**5** 9 and 303

**6** 21 and 35

**7** 2106 and 15

**8** 143 and 26

**9** 18 and 37

**10** 136 and 391

## Question 2

The integers less than 6 are 1, 2, 3, 4 and 5. Of these, only 1 and 5 are co-prime with 6. So there are 2 integers less than 6 that are co-prime with 6.

For every integer from 2–24 inclusive look at the numbers less than each that are co-prime with each.

We shall denote the number of integers less than $n$ and co-prime with $n$ as $\Phi(n)$.

Record your results in a table like this:

| integer | integers less than and co-prime with | number of such integers $\Phi(n)$ |
|---|---|---|
| 2 | | |
| 3 | | |
| 4 | | |
| 5 | 1, 2, 3, 4 | 4 |
| 6 | 1, 5 | 2 |
| 7 | | |
| etc | | |
| 24 | | |

# Question 3

**1** Does $\Phi(3) \times \Phi(4) = \Phi(12)$?     **2** Does $\Phi(2) \times \Phi(6) = \Phi(12)$?

In some cases

$$\Phi(n) \times \Phi(m) = \Phi(n \times m)$$

and in other cases this is not true. Experiment with a few choices of $n$ and $m$ for yourself and try to work out the circumstances under which it is true.

# Question 4

(i) For each of the prime numbers in your table, work out $\Phi(n)$.

(ii) If $p$ is *any* prime number state clearly the relationship which exists between $p$ and $\Phi(p)$.

# Question 5

If $p$ is any prime number and $a$ an integer, work out a formula for $\Phi(p^a)$.

*Hint*
Try $p = 2, 3, 5, 7, \ldots$ etc. and $a = 2, 3, 4, \ldots$ etc.
That is, look at:
$\Phi(2), \Phi(2^2), \Phi(2^3), \ldots$ etc. and $\Phi(3), \Phi(3^2), \Phi(3^3)$ etc.

Now try to find a pattern which can be **generalised**.

## Prime-factor form

We can work out the $\Phi$ value for a large number by putting that large number in **prime-factor** form and then using the result of question 3. You may have seen a method in chapter 29, The Sigma Function for finding prime-factors of a number.

> ## *Example*
>
> Divide the number repeatedly with prime numbers, starting with 2, until the result is no longer divisible by 2, then use 3, then 5, etc. through the other primes.

For 720 we have . . .

```
2)720
2)360
2)180
2)90
3)45
3)15
5)5
  1
```

So,

$$\Phi(720) = \Phi(2^4 \times 3^2 \times 5)$$
$$= \Phi(2^4) \times \Phi(3^2) \times \Phi(5)$$
$$= 8 \times 6 \times 4$$
$$= 192$$

## Question 6

Try to find the $\Phi$ values for

**1** 30                       **4** 420

**2** 60                       **5** 245

**3** 180                     **6** 1001

## A factors programme

This BASIC program works out, and prints the factors and the number of factors of a number *N*.

```
 10 PRINT "ENTER YOUR NUMBER"
 20 INPUT N
 30 LET P=1
 40 FOR K= INT(N/2) TO 1 STEP −1
 50 IF INT (N/K) <>N/K THEN 80
 60 PRINT K
 70 LET P=P+1
 80 NEXT K
 90 PRINT " THE NUMBER OF FACTORS IS. . . . . .";P
100 PRINT:PRINT:PRINT
110 GOTO 10
```

# Question 7

Can you write a program to determine

 (i) the common factors of 2 numbers

(ii) whether or not 2 numbers are co-prime?

   The $\Phi$ function was first investigated by Leonhard Euler (1707−1783). The function has been named after him and is now known as Euler's Function. It is very important in the study of Number Theory.

# 37 Areas Under Curves

In the late seventeenth and early eighteenth century, certain people were stimulated by the problem of trying to calculate areas of some unusual shapes. One of these problems was concerned with calculating the area under a curve.

The shaded region is the area we wish to calculate. This is the part of the graph $y = x^2$ between $x = 0$ and $x = 4$.

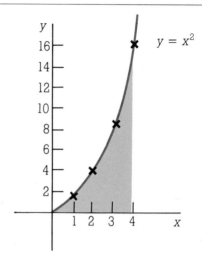

Before we have any technique to calculate the area, it is always useful to be able to estimate it. In this case the area of whole rectangle is

$$4 \times 16 = 64 \text{ square units}$$

The area under the curve is certainly less than half the area of the rectangle. Why?
   So the area under the curve is less than 32 square units. Therefore we can estimate that the area under the curve is likely to be somewhere in the region of 20 to 25 square units.

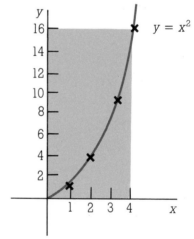

## Question 1

Obtain estimates for the areas under the following curves:

**1** $y = x^2$ as $x$ varies from 0 to 5

**2** $y = x^3$ as $x$ varies from 0 to 3

**3** $y = 2x^2$ as $x$ varies from 0 to 3

**4** $y = 2x^3$ as $x$ varies from 0 to 3

However mathematicians discovered better ways of estimating the area. Here are some.

## Using steps

*Upper steps*
The area under the curve is certainly less than the area of the upper steps.
Area of steps =

$$1 + 4 + 9 + 16$$

i.e.

$$1^2 + 2^2 + 3^2 + 4^2$$

$$= 30 \text{ square units}$$

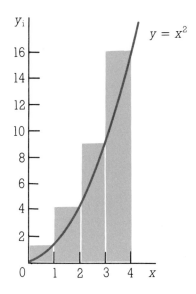

*Lower steps*
The area under the curve is also certainly greater than the area of the lower steps.
Area of steps =

$$0 + 1 + 4 + 9$$

i.e.

$$0^2 + 1^2 + 2^2 + 3^2$$

$$= 14 \text{ square units}.$$

The area under curve is less than the area of the upper steps and greater than the area of the lower steps. So if we let $A$ = area under curve

$$14 < A < 30$$

If the area is between 14 and 30, a reasonable estimate must be that it is the average of 14 and 30, which is 22.

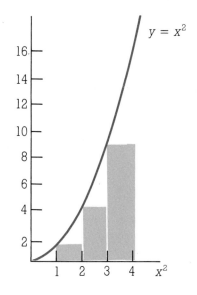

## Question 2

Using upper and lower steps, obtain estimates for the area under the curves

**1** $y = x^2$ as $x$ varies from 0 to 5

**2** $y = x^3$ as $x$ varies from 0 to 3

**3** $y = 2x^2$ as $x$ varies from 0 to 3

**4** $2x^3$ as $x$ varies from 0 to 3

## Question 3

Can you offer any suggestions for ideas in which we could, still using the steps, obtain even better estimates for the areas under the curves? Write your ideas down giving any diagrams or examples you feel are appropriate.

## Using Trapeziums

In this method we join up some points on the curve with straight lines. This creates a sequence of trapeziums, the areas of which approximate to the area under the curve.

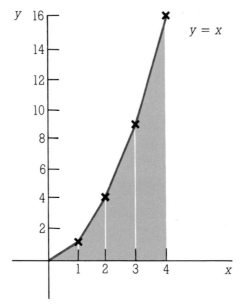

## Question 4

Show that the estimate of the area under the curve $y = x^2$ given by the trapeziums above is 22 square units.

## Question 5

Using trapeziums, obtain estimates for the areas under the curves

**1** $y = x^2$ as $x$ varies from 0 to 5      **3** $y = 2x^2$ as $x$ varies from 0 to 3

**2** $y = x^3$ as $x$ varies from 0 to 3      **4** $y = 2x^3$ as $x$ varies from 0 to 3

Compare these results with those obtained in questions 1 and 2.

## Accurate estimate

All graphs of the type

$y = x^n$ (where $n$ is a positive integer)

look something like this:

In this investigation you will be asked to look at some of these curves and the areas underneath them.

Given the value of $n$, it is possible to say that the area under the curve is a particular fraction of the area of the rectangle OABC. The particular fraction depends on the value of $n$.

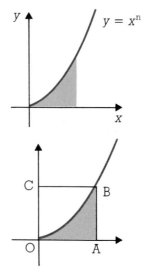

## Question 6

In order to find the relationship between the particular fraction and the value of $n$ look at the following graphs.

$$y = x$$
$$y = x^2$$
$$y = x^3$$
$$y = x^4$$

There is no need to look at values of $x$ any greater than $x = 3$.

There are many ways you could set about this investigation. Here are some examples.

---

### Example

(i) Draw the curves very accurately on graph paper and count the squares.

(ii) Draw and cut out the shapes of the area under the curve and the rectangle. Weigh these 2 shapes and find the fraction:

$$\frac{\text{weight of area under curve}}{\text{weight of area under rectangle}}$$

Because you can find the rectangle area accurately you can also find

---

the area under the curve accurately. However, you will need some sensitive weighing scales.

(iii) Obtain very good estimates through the techniques of questions 1–4.

(iv) Use a method of your own that you think is suitable.

How do you think the fraction

$$\frac{\text{area of the curve}}{\text{area of the rectangle}}$$

varies when the curve becomes $y = 2x^3$, $y = 3x^2$, $y = 5x$ or $y = 3x^6$? That is, what happens to the function in the **general case** when the curve is

$$y = ax^n$$

where $a$ is a constant?

## Question 7

Calculate the areas under the curves by a method of your choice

**1** $y = x^2$ as $x$ varies from 0 to 5      **6** $y = 2x^2$ as $x$ varies from 0 to 4

**2** $y = x^3$ as $x$ varies from 0 to 4      **7** $y = 3x^4$ as $x$ varies from 0 to 2

**3** $y = x^5$ as $x$ varies from 0 to 3      **8** $y = 2x^5$ as $x$ varies from 0 to 3

**4** $y = x^7$ as $x$ varies from 0 to 4      **9** $y = \frac{1}{2}x^3$ as $x$ varies from 0 to 4

**5** $y = x^{10}$ as $x$ varies from 0 to 2      **10** $y = 5x^2$ as $x$ varies from 0 to 3

## Further components

In the remaining questions the various curves have more than one component. We will illustrate the technique for dealing with them by looking at one very simple example.

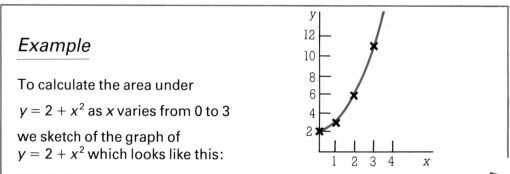

*Example*

To calculate the area under

$y = 2 + x^2$ as $x$ varies from 0 to 3

we sketch of the graph of
$y = 2 + x^2$ which looks like this:

The area underneath the curve can be broken into 2 parts:

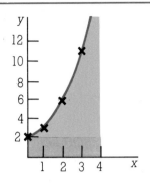

so:

$$\left[\begin{array}{c}\text{area under} \\ y = 2 + x^2\end{array}\right] = \left[\begin{array}{c}\text{area under} \\ y = 2\end{array}\right] + \left[\begin{array}{c}\text{area under} \\ y = x^2\end{array}\right]$$

and as $x$ varies from 0 to 3:

area under $y = 2$  is    $2 \times 3 = 6$ square units
area under $y = x^2$ is    $\frac{1}{3}$ area of rectangle
$= 9$ square units

So the area under $y = 2 + x^2$ as $x$ varies from 0 to 3 is 15 square units.

## Question 8

Calculate the area under each of the following curves:

**1** $y = 3 + x^2$ as $x$ varies from 0 to 4

**2** $y = 2 + x^3$ as $x$ varies from 0 to 2

**3** $y = 2x^2 + x^3$ as $x$ varies from 0 to 3

**4** $y = x + x^3$ as $x$ varies from 0 to 4

**5** $y = 2 + 4x + x^2$ as $x$ varies from 0 to 3

**6** $y = x^2 - 2x + 2$ as $x$ goes from 0 to 3

# 38 Gradient Function 1

The gradient of a curve or line tells us how steep that curve or line is. These four lines have different gradients. A straight line has a constant gradient.

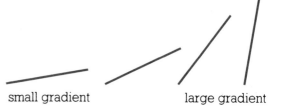

small gradient          large gradient

## Measuring the gradient of a straight line

### Example

Choose 2 appropriate points on the line and construct a right-angled triangle.

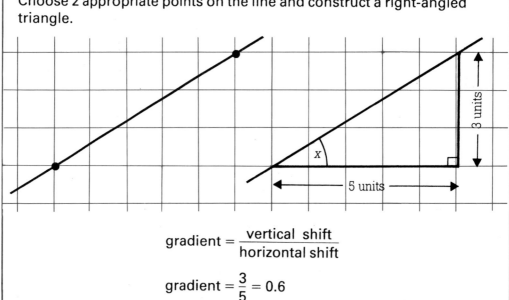

3 units

5 units

$$\text{gradient} = \frac{\text{vertical shift}}{\text{horizontal shift}}$$

$$\text{gradient} = \frac{3}{5} = 0.6$$

Note that

$$\text{tangent of angle } x = \frac{\text{opposite}}{\text{adjacent}} = \frac{3}{5}$$

Therefore the gradient of a straight line is the tangent of the angle that the line makes with the horizontal.

A line sloping this way has a positive gradient.

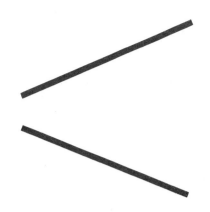

A line sloping this way has a negative gradient.

## Question 1

Calculate the gradients of each of these lines

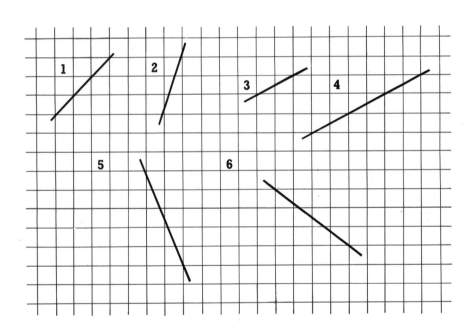

## Question 2

Calculate the gradient of the line which passes through the points

**1**  (3, 2) and (8, 9)

**2**  (4, 7) and (5, 9)

**3**  (−3, 2) and (8, 5)

**4**  (0, 6) and (4, 0)

**5**  (−2, 7) and (4, 1)

**6**  (0, −3) and (2, 4)

## Question 3

Draw an accurate diagram of lines with the following gradients:

| **1** 3 | **3** 1 | **5** 0 | **7** $-\frac{1}{3}$ | **9** $2\frac{1}{2}$ |
|---|---|---|---|---|
| **2** $-4$ | **4** $-1$ | **6** $\frac{1}{2}$ | **8** $-\frac{3}{5}$ | **10** $-3\frac{1}{3}$ |

## The gradient of a curve

These 3 curves are such that the first is not as steep as the second, which is not as steep as the third.

The diagrams below show how to measure the gradient at the point (2, 4) on the curve $y = x^2$

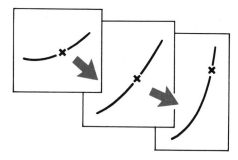

Draw the curve and mark the point (2, 4)

Draw the tangent to the curve at the point.

Construct the right-angled triangle

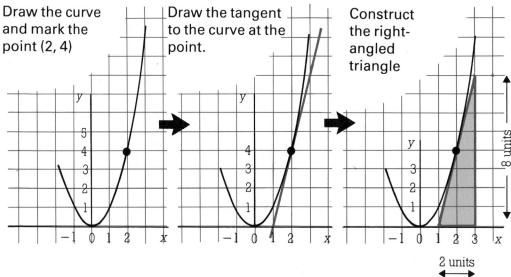

$$\text{Gradient} = \frac{y \text{ shift}}{x \text{ shift}}$$

$$= \frac{8}{2} = 4$$

The gradient of the curve at the point (2, 4) is 4.

## Question 4

Draw a very accurate graph of $y = x^2$, for $x = 1$ to $x = 4$. On the graph draw the tangents at the points: (1, 1) and (3, 9). Show that the gradient at (3, 9) is 6 and that the gradient at (1, 1) is 2.

## Question 5

We now have a few results relating the *gradient* to the *point* on the *curve*.

| curve | point | gradient |
|-------|-------|----------|
| $y = x^2$ | (1, 1)<br>(2, 4)<br>(3, 9) | 2<br>4<br>6 |

Find the gradients at these points

**1** (4, 16)  **3** (10, 100)  **5** (−5, 25)

**2** (5, 25)  **4** (−3, 9)  **6** (0, 0)

## The gradient function

The relationship between the gradient and the point on the curve can be **generalised** and is called the **gradient function**. The rest of this work is about answering the question:

Given the curve, what is its gradient function?

## Question 6

What do you think the generalised result is for the gradient function for the curve $y = x^2$?

## Alternative method

On the next page is a magnified sketch of the curve that was drawn earlier.

If the *nearby point* is very close to *the point* then the chord and the tangent are almost equal. So the gradient at the point on the curve is *almost equal* to the gradient of the chord.

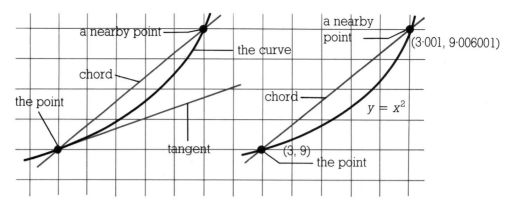

Let us see how this works out for the curve $y = x^2$ at the point (3, 9).

For the nearby point we have made a small increase of 0.001 in the $x$ value. The gradient of the chord

$$= \frac{(3.001)^2 - 3^2}{3.001 - 3} = \frac{9.006001 - 9}{.001} = \frac{.006001}{.001}$$

$= 6.001$ which is very nearly 6.

From our accurate construction we found that the gradient of the curve at (3, 9) was 6.

For the next questions you will need either a calculator or computer.

## Question 7

Use a small increase in $x$ of 0.001 to show that the gradient of the curve $y = x^3$ at the point (2, 8) is very nearly 12.

## Question 8

Again using a small increase in $x$ of 0.001, obtain the gradients

**1** at the point (2, 16) on the curve $y = x^4$

**2** at the point (2, 32) on the curve $y = x^5$

**3** at the point (2, 64) on the curve $y = x^6$

| curve | point | gradient |
|-------|-------|----------|
| $y = x^2$ | (2, 4) | 4 (= 2 × $2^1$) |
| $y = x^3$ | (2, 8) | 12 (= 3 × $2^2$) |
| $y = x^4$ | (2, 16) | |
| $y = x^5$ | (2, 32) | |
| $y = x^6$ | (2, 64) | |

Copy and complete the table above. Can you see a pattern in your results?

# Question 9

Using a small increase in $x$ of 0.0001, obtain the gradient as indicated below. The $x$ value is always 3.

| curve | point | gradient |
|-------|-------|----------|
| $y = x^2$ | (3, 9) | |
| $y = x^3$ | (3, 27) | |
| $y = x^4$ | (3, 81) | |
| $y = x^5$ | (3, 243) | |
| $y = x^6$ | (3, 729) | |

Do these follow any pattern you found in question 8?

# Question 10

Taking the points with an $x$ value of 4 and a small increase of 0.001, obtain the gradient on the curves

**1** $y = x^2$       **4** $y = x^5$

**2** $y = x^3$       **5** $y = x^6$

**3** $y = x^4$       **6** $y = x^7$

Does your pattern hold true?

# Question 11

Can you offer a **generalisation** for the gradient function for the curve: $y = x^n$? What are the gradients at:

**1** (2,1024) on the curve $y = x^{10}$       **3** (7,2401) on the curve $y = x^4$

**2** (3,2187) on the curve $y = x^7$       **4** (5,3125) on the curve $y = x^5$

# 39 Towers of Hanoi

There are 3 wooden towers $x$, $y$, $z$, with 7 discs of increasing radii stacked on tower $x$.

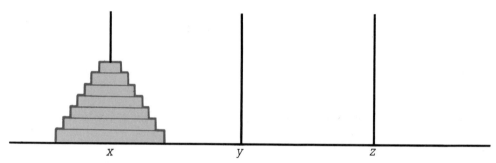

One disc at a time can be moved to any tower but you cannot place a larger disc on top of a smaller one. If you do not have the tower, you could use stacks of Cuisenaire rods, pieces of paper or numbered cards.

## Question 1

Can you move the 7 discs from tower $x$ to either $y$ or $z$ and finish in one of the following ways?

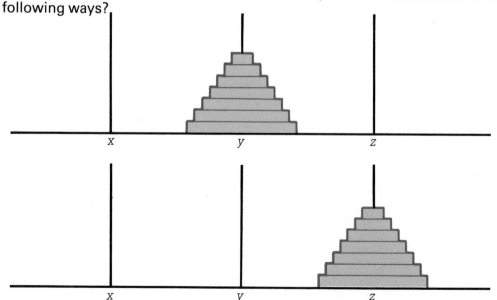

If you find it difficult with 7 discs, simplify the problem by first trying with 2 discs. Then try with 3, 4, 5, 6 and 7 discs. In each case try to complete the puzzle in the *least* number of moves. Record your results in a table like this:

| number of discs | 1 | 2 | 3 | 4 | 5 | 6 | 7 | 8 |
|---|---|---|---|---|---|---|---|---|
| number of moves taken | | | | | | | | |

What observations can you make about this table of results? Can you give a generalised result which connects the number of discs and the least number of moves?

## Question 2

What would be the least number of moves for

**1** 10 discs          **3** 100 discs

**2** 20 discs          **4** $n$ discs?

## Question 3

What would be the number of discs if the least number of moves was

**1** 511                      **2** 32767?

*Hint:* In working out the puzzle, you might find it helpful to count the number of moves made by each individual disc. Try it. What do you notice?

# 40 Billiards 2

## Rebounds

A billiard table with no centre pockets is in the form of a 3 × 5 rectangular grid. The ball always enters at the bottom right-hand corner and moves at 45° to the edges. It rebounds around the table before exiting at a corner.

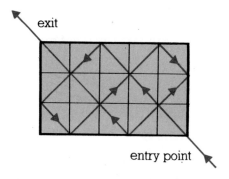

In this case the ball rebounds 6 times and exits at the top left hand corner.

. Your task is to determine the relationship between the grid size and the number of rebounds.

## Question 1

For each of the following grids draw the appropriate grid, show the path of the ball and count the number of rebounds.

**1** 4 × 7 grid        **3** 5 × 8 grid

**2** 3 × 8 grid        **4** 7 × 9 grid

## Question 2

Copy and complete the table below:

| grid size | number of rebounds |
|-----------|--------------------|
| 3 × 5 | 6 |
| 4 × 7 | |

continued . . .

| | |
|---|---|
| 5 × 9 | |
| 4 × 9 | |
| 8 × 11 | |
| 3 × 4 | |
| 5 × 7 | |
| 8 × 15 | |

In all of the cases examined so far the two numbers in the grid size are co-prime, that is their only common factor is 1.

# Question 3

What is the **general result** for a grid of size $n \times m$ with $n$ and $m$ co-prime? Give a formula for the number of rebounds.

# Question 4

Give a different grid size which will have the same number of rebounds as each of the following grids:

**1** 3 × 7          **2** 4 × 11          **3** 9 × 16

# Question 5

Examine the number of rebounds for grids of the following sizes:

**1** 3 × 6          **3** 8 × 10          **5** 4 × 8

**2** 6 × 9          **4** 6 × 15          **6** 4 × 6

# Question 6

What is the **general result** for the case where the grid numbers $n \times m$, have a common factor?

# Question 7

Can you offer a *proof* of your results, or some explanation of *why* the formula works?

## Question 8

For this 3 × 6 grid the ball does not pass through all the squares.

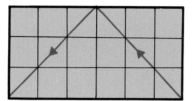

For this 3 × 5 grid the ball passes through all the squares.

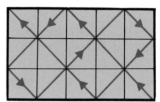

Under what circumstances does the ball pass through all the squares? **Generalise** your findings.

## Question 9

On the 2 × 5 grid, the ball is sometimes moving in an *anti-clockwise* sequence of moves.

However, sometimes it moves in a *clockwise* sequence.

On its route, the ball moves in an anticlockwise direction, then clockwise, then anticlockwise again thus making 2 changes in its direction of rotation. Investigate the relationship between the grid size and the number of changes in direction.

## Question 10

Investigate what happens if we change the angle of the ball's path.

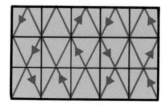

Can you find any general result for:

(i) the relationship between the grid sizes and the exit corner,

(ii) when the ball passes through all the squares,

(iii) the number of times the ball changes its direction of rotation.

## Question 11

Can you invent any billiards-type investigation of your own, based on a parallelogram-shaped table set out as an isometric grid?

# 41 Gradient Function 2

The gradient functions for the following curves are given below:

| curve | gradient function |
|---|---|
| $y = x^2$ | GF $= 2x$ |
| $y = x^3$ | GF $= 3x^2$ |
| $y = x^4$ | GF $= 4x^3$ |
| $y = x^5$ | GF $= 5x^4$ |
| $y = x^6$ | GF $= 6x^5$ |

We shall write this as, for example:

$$\text{GF}(x^3) = 3x^2 \text{ or in \textbf{general} terms: GF}(x^n) = nx^{n-1}$$

How does this work out for the straight line $y = x$?

## Question 1

Refer to the techniques used in Chapter 38 before attempting these questions. You must justify your answers in each case.

**1** Does GF $(x^2 + x^3) =$ GF $(x^2) +$ GF $(x^3)$?

**2** Does GF $(2x^3) = 2 \times$ GF $(x^3)$?

**3** Does GF $(x^2 + x^5) =$ GF $(x^2) +$ GF $(x^5)$?

**4** Does GF $(x^n + x^m) =$ GF $(x^n) +$ GF $(x^m)$?

**5** Does GF $(k\,x^n) = k\,$GF $(x^n)$

**6** Does GF $(x^n - x^m) =$ GF $(x^n) -$ GF $(x^m)$?

## Question 2

Calculate the gradients of the curves by substituting the $x$ value into the gradient function.

**1** $y = x^3$ at the point where $x = 4$

**2** $y = 3x^5$ at the point where $x = 3$

**3** $y = 2x^2 - 3x$ at the point where $x = 3$

**4** $y = x^4 + 2x^2 + 4x$ at the point where $x = 1$

## The Lines $y$ = constant

The graph shows the lines $y = 5$ and $y = 3$.

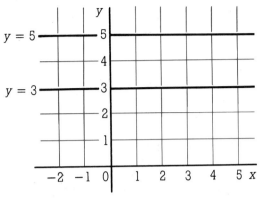

Any line where the equation has

$$y = \text{constant}$$

is horizontal and hence its gradient is 0. So if we have the lines

| | |
|---|---|
| $y = 3$ | GF = 0 |
| $y = 5$ | GF = 0 |
| $y = \text{constant}$ | GF = 0 |

## Question 3

Write down the gradient function for the curve $y = x^3 + 5x^2 + 3x - 7$. Hence calculate the gradient at the point where $x = 2$.

Suppose we want to know the co-ordinates of a particular point on a curve but only know the gradient at that point and the equation of the curve.

---

### Example

If the curve is $y = x^2 - 3x + 2$ and if the gradient at a point is 5 then we can find the co-ordinates like this:

$$y = x^2 - 3x + 2$$
$$GF = 2x - 3$$

At the required point the gradient is 5. So

$$2x - 3 = 5$$
$$2x = 8$$
$$x = 4$$

When

$$x = 4$$
$$y = x^2 - 3x + 2$$
$$y = 4^2 - 3(4) + 2$$
$$y = 6$$

So the co-ordinates of the required points are (4, 6).

---

## Question 4

You will now be given some curves and the values of the gradient at a point. You have to find the co-ordinates of the points. Be careful with 4 and 5.

**1** $y = x^2 - 7x + 3$, gradient $= 1$      **4** $y = x^3 + 7x$, gradient $= 16$

**2** $y = 3x^2 - 16x + 15$, gradient $= 2$      **5** $y = 3x^2 - 18x + 25$, gradient $= 3$

**3** $y = x^2 + 5x - 2$, gradient $= -3$

## Turning points

At turning points the gradient is 0 because the tangent line is horizontal.

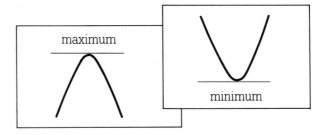

maximum

minimum

---

*Example*

Calculate the co-ordinates of the turning point on the curve $x^2 - 6x + 2$.

If
$$y = x^2 - 6x + 2$$
$$GF = 2x - 6$$

At the turning point, GF $= 0$

$$2x - 6 = 0$$
$$2x = 6$$
$$x = 3$$

when $x = 3$

$$y = x^2 - 6x + 2$$
$$y = 3^2 - 6(3) + 2$$
$$y = 9 - 18 + 2$$
$$y = -7$$

So the co-ordinates of the turning point are $(3, -7)$

---

# Question 5

Calculate the co-ordinates of the turning points on the curves

**1** $y = x^2 - 8x + 9$            **3** $y = 2x^2 - 12x + 17$

**2** $y = 4 - 10x + x^2$        **4** $y = 5 + 6x - x^2$

In the example, we found a turning point at $(3, -7)$. If we look either side of $x = 3$ when $x = 2$ and $x = 4$, we have:

$$\begin{aligned} y &= x^2 - 6x + 2 & y &= x^2 - 6x + 2 \\ &= 2^2 - 6(2) + 2 & &= 4^2 - 6(4) + 2 \\ &= 4 - 12 + 2 & &= 14 - 24 + 2 \\ &= -6 & &= -6 \end{aligned}$$

So we have a sketch like this which shows a minimum turning point.

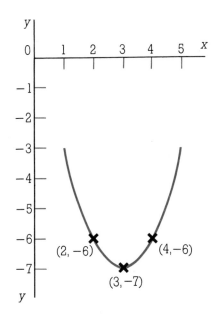

# Question 6

(i) Work out whether the turning points in question 5 are maximum or minimum ones.

(ii) Calculate the co-ordinates of the turning points on the curve

$$y = x^3 - 9x^2 + 24x + 4$$

Decide whether the turning points are maximum or minimum.

# 42 Scores

## Half-time scores

The hockey match between Wickham Wanderers and Clarbury All Stars finished with Wickham winning 3–1. When 2 young supporters saw this result they wondered what the half-time score might have been. They worked out the following possibilities:

3–1, 3–0, 2–0, 2–1, 1–0, 0–0, 0–1, 1–1.

They noticed that there were 8 different, possible half-time scores. This made them wonder what the relationship was between the *final score* and the number of *possible half-time scores*.

## Question 1

Work out the possible half-time scores for matches that end with final scores of:

**1** 1–1                    **4** 3–2

**2** 2–1                    **5** 4–2

**3** 4–1                    **6** 0–0

Do the same for 3 final scores of your own choice.
  Record your results, preferably as a table showing the final score and the number of possible half-time scores. Write down anything you notice about the result and try to **generalise** it.

## Question 2

Use your working from question 1 to calculate the number of possible half-time scores in matches where the final score is:

**1** 4–3                    **3** 8–7

**2** 5–5                    **4** 13–2

## Score sequences

When the final score is 3–1, there are different sequences of part-time scores leading to it. Here is 1 example:

$$(0-0) \rightarrow (1-0) \rightarrow (2-0) \rightarrow (3-0) \rightarrow (3-1)$$

## Question 3

There are 3 other sequences of results leading to a final score of 3–1. Write them down.

The total number of sequences leading to 3–1 is 4. Take some other score and look at the number of possible sequences leading to that result. Try to **generalise** the result but it will not be easy.

# 43 Sticks

Here are 2 pictures of 3 lines; each line could go on forever (be of infinite length).

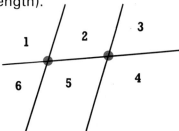 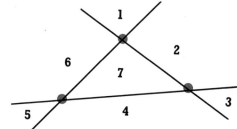

The second picture shows the maximum number of **crossover points** and the maximum number of **regions** for an arrangement of 3 straight lines.

Your task is to look at the relationship between the number of lines and the maximum number of crossover points and regions.

## Question 1

Draw the pictures for the maximum number of crossover points and regions for:

**1** 2 lines        **2** 4 lines        **3** 5 lines

## Question 2

What are the results for the numbers of crossover points and regions when you have only 1 line? What about when you have 0 lines?

## Question 3

Can you use your results to work out the maximum numbers of crossover points and regions for large numbers of lines. That is, can you generalise your results? Can you predict the maximum numbers of *crossover points* and *regions* with 20 lines and 100 lines?

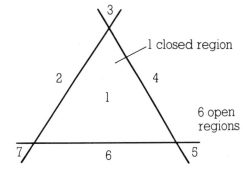

## Question 4

Can you find and prove any **general relationship** between the number of lines and the numbers of open regions and closed regions?

## Question 5

The diagram below shows how to make the maximum number of cross-over points with 5 lines. The enclosed regions have been shaded. There are 3 quadrilaterals and 3 triangles.

Is there a **general relationship** between the types and numbers of enclosed regions and the numbers of lines?

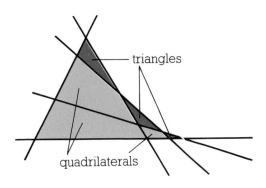

## Question 6

The crossover points and regions could be made by things other than lines, for instance, circles. Examine the relationship between the number of circles and the numbers of crossovers and regions.

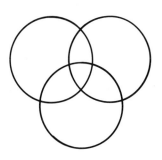

## Question 7

Examine cases where you mix lines and circles, for instance 2 circles and 1 line gives 4 crossover points and 8 regions.

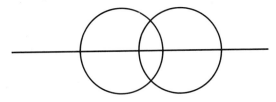

# Question 8

Examine the maximum number of lines and regions formed by intersecting planes. For instance, in 3 dimensions, 2 planes give:

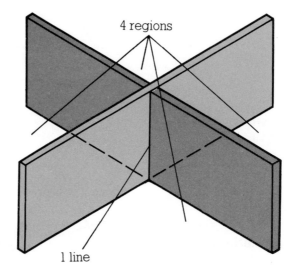

4 regions

1 line

What about 3 planes, 4 planes, 5 planes?
   What is the maximum number of regions obtained with 5 planes?
   Can you find a general result?

# 44 3-D Permutations

When looking at the different ways 2 identically-sized but different coloured cubes can be placed together, we need the following rules:

(i) Full faces must be used so that this is allowed

but neither of these are allowed.

(ii) We consider positions 'inside' a 3-dimensional set of axes, with a cube placed in the 'home' position (as if at the bottom inside corner of an open shoe-box) so that, for 3 cubes, this is allowed:

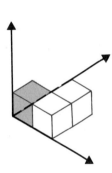

but these are not:

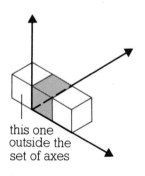

this one outside the set of axes

*or*

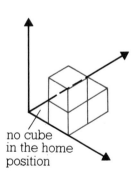

no cube in the home position

*or*

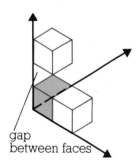

gap between faces

## Question 1

Check that the following are the only 6 allowable positions for 2 cubes of different colours. R = red; B = blue.

175

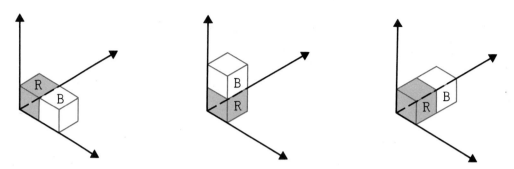

Interchange the position of the red and the blue to find the other 3 positions.

## Question 2

Use diagrams, with a detailed explanation, to investigate the number of different positions or **permutations** for 3 cubes under the same conditions.

Two examples of allowable permutations for red, blue and green cubes are:

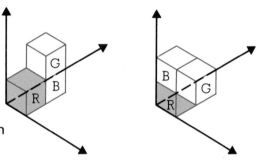

## Question 3

How many permutations are there for 3 cubes when

(i) all 3 cubes are the same colour

(ii) 2 of the cubes are the same colour?

Extension problem

## Question 4

How many permutations are there for 4 cubes when

(i) all 4 cubes are the same colour

(ii) all 4 cubes are different colours

(iii) some cubes are the same colour and some are different?